Vollers' A.-F. Code

Anglo-French
General Telegraphic Code

Code Général Télégraphique
Anglo-Français

compiled and published by
compilé et édité par

R. VOLLERS
Munich

Printed by / *Imprimé par* — E. Mühlthaler's Buch- u. Kunstdruckerei A.G., Munich

1904

Price including Mutilation key:
Prix inclus Clef des mutilations:

30/— net cash or / *ou* *frs. 38.— net comptant.*

volcaba	= read the following code cipher from this Code
volcad	= ,, ,, ,, 2 code ciphers ,, ,, ,,
volcadolo	= ,, ,, ,, 3 ,, ,, ,, ,, ,,
volcadome	= ,, ,, ,, 4 ,, ,, ,, ,, ,,
volcale	= ,, ,, ,, 5 ,, ,, ,, ,, ,,
volcamonos	= ,, all the following code ciphers ,, ,, ,,
volcamos	=
volcandola	=
volcandote	=
volcanlo	=
volcarais	= traduisez le chiffre suivant par ce Code
volcaralo	= ,, les 2 chiffres suivants ,, ,, ,,
volcarela	= ,, ,, 3 ,, ,, ,, ,, ,,
volcarian	= ,, ,, 4 ,, ,, ,, ,, ,,
volcarlos	= ,, ,, 5 ,, ,, ,, ,, ,,
volcaseis	= ,, tous les chiffres suivants ,, ,, ,,
volcola	=
volcoles	=
volcose	=

NB. The above code ciphers are Spanish words.
Les chiffres de code ci-dessus sont des mots espagnols.

Preface.

Important remarks for the use of this Code.

The matter of this code is divided into 2 sections:

1) The **"Sundries" (Divers) Section,** pages 1 to 49 based on the ordinary system needing no explanation, and

2) The **Section of Tables** 1—36 based upon a system of multiplication i.e. a combination of roots placed on the left hand side, marked ■—►, with the corresponding terminations marked ◄—■ on the right hand side **of the same table.**

The roots have a hyphen (-) appended to them, whereas the terminations have this hyphen prefixed to them; each root can be prefixed to each termination of the **same** table forming thus the whole word to be transmitted by the Telegraph Office.

The **Index of Roots** serves to refer the translator of the telegram to the pages or tables where the ciphers (words) are to be found.

Read the groups:

	text	text
text text		

from below to above and those:

text	text	
		text text

from above to below.

The **guiding lines** after each 5th sentence serve to facilitate the reading of the text from right to left and vice-versa.

The sign 〰 means "for general use" f. i. denot- (table 9) could not be wired alone without being prefixed to a termination, as it would not represent a complete word and would therefore not correspond with the rules for international telegrams.

The double text of the tables is composed with **absolute identity of meaning.**

If however a firm should like to mention the language in which it is wiring it would be sufficient to employ one of the words: volcaba etc. page I or one from p. 1—49 which have not the double text.

Préface.

Notes importantes pour l'usage de ce Code.

Les matières de ce code sont réparties en 2 sections:

1) La **Section « Divers »** (Sundries) pages 1—49 basée sur le système ordinaire, qui n'exige aucune explication.

2) La **Section des Tables** 1—36 basée sur un système de multiplication, c. à d. une combinaison de syllabes radicales placées à la page gauche, marquées ■—►, avec les terminaisons correspondantes marquées ◄—■ placées à la page droite **de la même table.**

Les syllabes radicales sont suivies d'un trait (-) tandis que les terminaisons en sont précédées. Chaque radical d'une **même** table peut être combiné avec chaque terminaison de cette même table, formant ainsi le mot entier qui se transmet par les Bureaux des Télégraphes.

l'Index of Roots (syllabes radicales) s'emploie pour la traduction des télégrammes et indique les pages ou tables où les chiffres (mots) télégraphiés se trouvent.

Lisez les groupes:

	texte	texte
texte texte		

de bas en haut et les groupes:

texte	texte	
		texte texte

de haut en bas.

Les **lignes-guides** horizontales après chaque cinquième phrase servent à faciliter la lecture du texte de droite à gauche et vice-versa.

Le signe 〰 signifie: usage général, p. exemple: denot- (table 9) ne pourrait être télégraphié seul sans être combiné avec une terminaison, car il ne représenterait pas un mot entier et ne correspondrait alors pas aux règlements du service télégraphique international.

Le double texte des tables présente une **identité absolue de signification.**

Si toutefois une maison tenait à mentionner la langue en laquelle elle télégraphie, il lui suffirait d'employer un des mots volcaba etc page I ou bien ceux des pages 1 à 49. n'ayant pas le double texte.

Advantages of my system.

1. **Enormous reduction of volume**: about 85.000 phrases or terms of each of the 2 languages, viz.: a total of about 170.000 phrases or terms distributed on about 110 pages only, whereas according to the ordinary system about 3000 pages would have been necessary to express the same matter.

2. **Enormous economy of time** in composing and translating telegrams. The whole is arranged as far as possible alphabetically, even the English headings.

3. **Danger of the ciphers** (words) arriving mutilated is reduced to a minimum as the ciphers (words) chosen belong to the Latin, Spanish and Italian languages, representing a most harmonious alternation between consonants and vowels, so that the telegraphist will not meet with the same difficulty for their transmission which the use of words from the German languages would cause him.

4. **Facility in tracing mutilations** by means of the mutilation key containing the alphabetical arrangement of the terminations and roots by their **last** letter.

5. **A wrong interpretation** by mutilation or illegible and incorrect writing of the ciphers (words) almost **impossible** as they differ from each other by at least 2 characters; thus ciphers (words) such as "volare" and "violare" have been avoided.

Remarks about the choice of the ciphers.

The ciphers employed represent forms of conjugation of **active** (transitive) verbs of which:—

the **Latin** verbs are extracted from the Latin-German Dictionary by Karl Ernst Georges, Leipzig 1879/80,

the **Italian** verbs from the Dictionaries: Italian-French by Ferrari-Caccia and Italian-German by H. Michaelis, Leipzig 1895

and the **Spanish** verbs from the Spanish-French Dictionary by Fernandez Cuesta, Barcelona 1885.

These words are in strict conformity with the rules for international telegrams in force **at present.**

Avantages de mon système.

1. **Réduction énorme de volume:** env. 85.000 phrases ou termes de chacune des 2 langues, soit un total d'environ 170.000 phrases ou termes distribués sur environ 110 pages, tandis que d'après le système ordinaire il aurait fallu environ 3000 pages pour exprimer la même matière.

2. **Economie énorme de temps** pour la composition et la traduction des télégrammes. Le tout est rangé autant que possible par ordre alphabétique, même les chapitres anglais.

3. **Danger de mutilation** des chiffres (mots) est réduit à un minimum, les chiffres (mots) choisis appartenant aux langues latine, espagnole et italienne, présentant une alternation des plus harmonieuses entre consonnes et voyelles, de sorte que le télégraphiste n'éprouvera pas les difficultés pour leur transmission que lui provoquerait l'emploi des mots tirés des langues germaniques.

4. **Facilité de rechercher les mutilations** éventuelles au moyen de la clef des mutilations contenant le groupement alphabétique des terminaisons et des radicaux par la **dernière** lettre.

5. **Fausse interprétation par** suite de mutilation **ou** écriture illisible et inexacte des chiffres (mots) est presque tout-à-fait **impossible**, attendu qu'ils diffèrent entre eux par au moins 2 caractères. Ainsi les mots tels que « volare » et « violare » ont été évités.

Notes sur le choix des chiffres (mots).

Les chiffres employés représentent des formes de conjugaison de verbes **actifs** (transitifs) dont:

les verbes **latins** ont été extraits du dictionnaire Latin-Allemand par Karl Ernst Georges, Leipzig 1879/80,

les verbes **italiens** des dictionnaires Italien-Français par Ferrari-Caccia et Italien-Allemand par H. Michaelis, Leipzig 1895

et les verbes **espagnols** du dictionnaire Espagnol-Français par Fernandez Cuesta, Barcelone 1885.

Ces mots correspondent strictement aux règlements du Service télégraphique international **en vigueur à présent.**

The number of terminations has been increased (where this was grammatically admissible) by appending personal pronouns to certain Italian and Spanish terminations and "ne" (=of it, of them) to certain Italian terminations.

In the Berne Vocabulary, 2nd ed. ample use has been made of this appending of pronouns to forms of **active** verbs.

Munich, 10th February 1904.

Le nombre des terminaisons italiennes et espagnoles a été augmenté (où ceci était grammaticalement admissible) par l'addition de pronoms personnels et en outre de « ne » (= en) à certaines terminaisons italiennes.

Il est du reste fait abondamment usage de cette addition de pronoms à certaines formes de verbes a c t i f s, dans le Vocabulaire de Berne 2e édition.

Munich, le 10 Février 1904.

Example of a telegram

avvilammo	=	the firm (person) in question is establ. as importer/s, is said to be very competent, might be considered good for credit to the extent mentioned
pariabaris	=	shall act as proposed in your telegr. 5th curr. m.
amylabare	=	completed all contracts
adojadolo	=	as per your last telegr. we (I) accept offer expressed by the 14th word
hincarelas	=	sold 15000 yards
balteabo	=	$4^5/_{32}$ d.
sentitala	=	net per yard cif your port net 3 mo or $2^1/_2$ % 30 days

The cipher "amylabare" indicates that the telegram is to be read in English.

Exemple d'un télégramme

volcaseis	=	traduisez tous les chiffres suivants par ce Code
Muller	=	Muller
dijugantis	=	Fils & Co. de votre ville télégraphiez renseignements et si cette maison pourrait être considérée bonne pour crédit de quelques milliers de francs
spicatae	=	faites au mieux; ceci en réponse à votre lettre du 9 p.
oculabo	=	marchandise commandée s'expédiera le 10 ct.
spurcares	=	remis 7750 frs
gravemus	=	à valoir sur encaissement

Les chiffres: « volcaseis » et « gravemus » indiquent que ce télégramme doit être lu en français, du reste l'un de ces 2 chiffres seul suffirait à cet effet.

Table of Contents. *Table des Matières.*

p. 1—23	Sundries (English Section)	45 roots *radicaux*	× 55 termns	= 2475	Latin ciph.	*chiffres latins*
„ 24—49	Divers *(Sect. française)*	51 „	× 55 „	= 2805	„ „	„ „
Table 1	Correspondence and Telegrams	58 „	× 55 „	= 3190	„ „	„ „
„ 2	Telegrams	21 „	× 56 „	= 1176	Spanish „	„ *espagn.*
„ 3	Dates	72 „	× 31 „	= 2232	Latin „	„ *latins*
„ 4	„	34 „	× 39 „	= 1326	„ „	„ „
„ 5	Epochs *(Epoques)*	14 „	× 83 „	= 1162	Italian „	„ *italiens*
„ 6	„ „	48 „	× 25 „	= 1200	Latin „	„ *latins*
„ 7	Hours *(Heures)*	38 „	× 28 „	= 1064	„ „	„ „
„ 8	„ „	24 „	× 60 „	= 1440	Italian „	„ *italiens*
„ 9	Asking for Information *(Demandes de renseignements)*	88 „	× 37 „	= 3256	Latin „	„ *latins*
„ 10	Giving Information *(renseignements fournis)*	133 „	× 62 „	= 8246	Italian „	„ *italiens*
„ 11	Finance	57 „	× 27 „	= 1539	Latin „	„ *latins*
„ 12	„	54 „	× 37 „	= 1998	„ „	„
„ 13	Movements of market &c. *(mouvements du marché etc.)*	20 „	× 78 „	= 1560	Italian „	„ *italiens*
„ 14	Movements of demand &c. *(mouvements de demande etc.)*	23 „	× 58 „	= 1334	„ „	„ „
„ 15	Statistics	9 „	× 39 „	= 351	Latin „	„ *latins*
„ 16	„	34 „	× 53 „	= 1802	Italian „	„ *italiens*
„ 17	Operations, Quantities	16 „	× 56 „	= 896	Spanish „	„ *espagn.*
„ 18	„ „	16 „	× 60 „	= 960	Italian „	„ *italiens*
„ 19	„ „	16 „	× 60 „	= 960	„ „	„ „
„ 20	„ Phrases	54 „	× 65 „	= 3510	Spanish „	„ *espagn.*
„ 21	„ „	16 „	× 65 „	= 1040	„ „	„ „
„ 22	„ „	16 „	× 65 „	= 1040	„ „	„ „
„ 23	±	23 „	× 83 „	= 1909	Italian „	„ *italiens*
„ 24	Prices, Amounts *(Prix, Montants)*	63 „	× 28 „	= 1764	Latin „	„ *latins*
„ 25	idem	100 „	× 80 „	= 8000	Italian „	„ *italiens*
„ 26	Prices, Quantities, Amounts *(Prix, Quantités, Montants)*	101 „	× 65 „	= 6565	„ „	„ „
„ 27	Quantities, Amounts *(Quantités, Montants)* 100—9950	91 „	× 56 „	= 5096	Spanish „	„ *espagn.*
„ 28	idem 10,000—10,050000	92 „	× 47 „	= 4324	„ „	„ „
„ 29	General Sentences *(Phrases Générales)*	50 „	× 65 „	= 3250	„ „	„ „
„ 30	idem	18 „	× 65 „	= 1170	„ „	„ „
„ 31	Shipments *(Expéditions)*	24 „	× 64 „	= 1536	„ „	„ „
„ 32	Ports &c.	15 „	× 65 „	= 975	„ „	„ „
„ 33	„ „	15 „	× 65 „	= 975	„ „	„ „
„ 34	Terms *(Conditions)*	50 „	× 62 „	= 3010	Italian „	„ *italiens*
„ 35	Private *(Privée)*	41 „	× 37 „	= 1517	Latin „	„ *latins*
„ 36	„ „	172 „	× 47 „	= 8084	Spanish „	„ *espagn.*
				total 94737	ciphers	*(chiffres).*

Index of Roots *(Syllabes Radicales)*

the figures not preceded by a " p " are the table numbers

les chiffres sans la lettre «p» indiquent les nos des tables

A

abaj 20
bon 10
buj 10
dic 35
ebr 28
em 3
ige 24
ism 28
iss 10
it 26
jur 10
leg 35
loc 35
nod 35
ol 34
on 20
ond 10
ord 28
ort 10
ram 10
rev 28
rom 28
rut 28
sum 6
ult 28
ur 20
us 5
accas 10
cin 6
cit 24
cod 10
cud 6
cul 10
cur 35
ed 20
ens 28
ept 28
er 2
erb 10
err 28
esm 10
et 5
hin 28
ich 28
it 20
lam 28
opl 28
os 20
qu 5
tiv 28
tu 20
u 3
acuci 28
un 2
ut 34
adaer 35
am p. 1
apt 35
aqu 10
arv 28
ast 10
at 2
bib 6
dit 10
dom 10
duc 6
ecu 28
ed 3
egu 10
em 20
eud 28
hal 24
iet 28
ig 3
im 26
ind 6
ir 5
izz 10
jug 35
jur 28
juv 10
lev 24
mir 28
ob 20
oj 2
on 5
or p. 1
orb 10
ot 26
rad 6
uan 28
ub 2
ugn 10
uj 20
ul 5
un p. 2
ur 3
us 26
vel 24
ver 28
aemul 24
qu p. 2
r p. 3
rm 2
stu 24
afam 2
afect 28
elp 28
fam 10
fid 10
fl p. 3
fun 10
ibl 28
il 2
in 20
loj 28
od 2
of 20
ond 28
um 20
agach 28
azz 10
enz 10
gel I0
ger 35
gir 10
it p. 4
lob 28
obi 28
ogn 10
ol 2
ram 28
ri 2
ris 28
rup 28
uij 28
uj 2
un 5
uz 17
ahetr 28
ij 20
irm 28
it 2
og 17
orn 28
oy 2
uch 28
ull 28
us 20
ajeb 2
egu 28
en 20
or 20
ust 28
ut 5
alaci 28
adr 28
amp 28
ap 4
arm 28
alaud 28
ber 10
bic 24
egr 28
ej 20
ert 28
gam 28
id 34
ien 35
if 20
ij 2
ind 28
ipi 28
ist 28
lab 24
lid 6
lig 35
lir 10
lu 3
men 28
od 20
oit 28
om 2
t p. 4
un 20
z 26
amain 28
ant 28
ar 4
as 20
att 10
bed 6
big 6
br 5
bul 35
bur 6
en 5
er 20
ic 4
im 20
maj 10
mir 10
mod 10
oen 24
ort 28
put 35
yl p. 5
anax 4
cl p. 5
ell 10
ex 20
gul 24
hel 35
hil 28
anill 28
im p. 6
n p. 6
nev 10
not 35
nu 34
nus 10
ot 2
ubl 28
ud 20
ul 2
apac 17
and 28
art 28
e 20
ell 28
er 34
est 28
ip 20
lan 28
oy 20
paj 10
par 35
ret 28
ro 20
unt 28
up 2
aquej 28
arbor 10
cu 20
din 28
el 20
gan 10
gil 28
gu 3
gur 28
gut 24
iet 35
m p. 7
nes 10
pon 28
rap 28
red 10
rid 28
rig 6
riv 10
rod 6
rog 35
roj 28
rol 10
rub 10
ruf 28
tu 4
uf 2
asalt 28
cus 28
es 27
ign 28
il 20
in 27
ob 27
oci 28
ol 5
orb 28
pas 28
per 10
pir 35
ser 6
sit 10
sol 24
son 10
um 27
atabl 28
air 28
edi 28
enu 28
et 20
igr 28
ild 28
ipl 28
iv 27
och 28
op 20
or 27
tel 10
ter 6
tig 6
tiv 10
ton 24
tu 5
uf 20
us 27
aucup 35
dit 24
gur 35
pr 20
rig 35
tum 35
avall 10
an 20
anz 10
eh 3
ell 6
en 27
err 6
is 27
isp 36
iv 20

the first letter of a termination is always a vowel — *la 1ère lettre d'une terminaison est toujours une voyelle*

avoc p. 7
ver 10
vil 10
azem 27
olv 36
orr 36
ot 20
ufr 36
ul 20
zim 10

B

bagn 5
j 26
jul 10
ld 20
li 27
lte 24
raj 36
rd 26
rlo 36
rr 5
si p. 8
toj 36
ttu 6
ub 4
bebd 20
f 20
ff 26
ix 20
ld 27
ll 4
nd 13
tun 36
biad 13
b 3
ff 13
n 27
ond 10
rl 20
tum 10
ul 26
zm 20
blasm 36
ind 10
boch 20
let 36
ll 13
nd 27
rd 13
rn 27
rr 20
sl 20
ved 36
y 20
zz 13
bram 26
as 20
av 13
evi 35
id 27
ind 36
om 27
ot 20
um 20
un 34
buh 21
lin 10
ll 21
ril 36
rl 13
rx 21
s 26
sc 17
ss 13
tt 13

C

cabs 27
c p. 8
ed 3
el p. 9
lce 35
lig 35
ll 21
lm 13
lon 36
lv 27
mbi 36
mel 36
mer 24
mod 36
mps 24
nd 21
ns 13
nt p. 9
pon 10
pp 13
pri 24
pt 21
pul 24
rat 10
rd 27
rem 36
ren 10
rin 24
rp 34
str 35
tt 13
us 21
ut 26
vil 36
z 17
zz 13
ceb 27
cin 36
dr 13
ff 13
ler 35
nn 13
nsu 10
pp 26
rc 17
rnu 24
rr 26
rt 4
st 34
chaf 21
ant 36
arp 36
as 27
et 13
iod 10
ipr 36
om 21
or 27
orr 36
um 27
up 21
urn 36
cib 26
cl 21
cur 10
fr 13
gn 23
ll 3
m 23
ng 3
nt 21
t p. 10
uff 10
urm 10
v 34
clam 23
ar p. 10
ass 10
av 21
ep 3
in 4
ipe 35
is 21
od 3
coact 24
add 6
al 3
bij 36
br 21
ed 21
em 3
en p. 11
ept 35
git 36
hob 36
coir p. 11
ll 26
lm 22
lor 35
lp 23
ltr 10
mb 22
med 6
mis 36
mit 35
mme 35
mod 10
mp 34
mpr 10
mun 10
nd 34
nfl 35
nlo 36
nsu 6
ort 10
pel 36
pin 36
qu 3
rch 36
rl 22
rn 23
ron 35
rru 6
rt 27
vid 10
vr 34
yt 22
crax p. 12
em p. 12
ep 23
ib 22
ibr 35
in 27
ism 36
isp 24
ost 10
uci 35
ust 35
uz 17
cuaj 22
art 36
b 22
ch 27
cin 10
di 27
est 36
id 22
it 23
ll 23
lp p. 13
nct 24
ne p. 13
r 26
rr 23
curs 22
rv p. 14
sc 34

D

dabi 22
div 36
m 26
mn p. 14
n 22
nn 23
pin 24
t p. 15
dealb 35
am p. 15
aur 35
can 6
cid 6
cim 10
cor 10
dom 24
flu 6
gn 23
gul 24
gun 6
hes 36
j 22
lav 24
lax 36
lib 10
lig 35
lin 6
met 9
mor 36
mud 36
neg 9
nod 10
not 9
ns p. 16
occ 24
opt 24
pil 9
pon 6
pos 36
ps 3
pur 10
put 9
puv 6
rad 6
riv 9
rog 9
ru 3
sal 10
sec 24
sin 10
soj 36
sol 9
desud 9
sug 6
sun 36
teg 6
ton 10
tt 26
ung 6
van 36
ved 36
vel 24
vis 36
vit 9
vor 9
zm 22
dibuj 36
ct p. 16
fam 36
ffl 9
fil 10
gn p. 17
jug 9
lat 9
lu 3
moj 10
not 10
nud 10
pan 10
pel 10
put 10
ren 10
rim 6
rob 10
rup 10
sat 10
sc 3
sen 36
sip 36
sir 10
sus 10
tt 23
vex 24
vid 6
vin 10
vor 10
dobl 26
l 27
lc 34
men 36
n p. 17
s 26
drag 17
ans 36
iz 17
dubd 22
ell 10
p 22
pl 4
r p. 18

E

ebri p. 18
eccit 10
edent 9
ic 3
ol p. 19
r 22
uc p. 19
effet 24
flu 6
foc 24
fug 9
egel 4
er 3
u 29
ejer p. 20
elect 24
ev p. 20
ij 29
ing 6
ix p. 21
oc p. 21
ogi 9
ong 24
ud 3
eman 26
anc 24
baf 36
bij 36
bud 36
end 9
erg 6
et 3
igr 9
ol 3
paj 36
pil 36
pop 36
pun 36
ung 6
usc 24
ut p. 22
enagu 36
alb 36
arr 10
art 36
asp 36
at p. 22
caj 36
cep 36
cim 36
don 36
ej 29
erv 9
fil 36
gaf 36
gib 36
engom 36
had 36
hen 36
hot 36
jeb 36
jut 36
lam 36
los 36
oj 27
ri 29
run 36
san 36
sop 36
tes 36
tr 29
ud p. 23
vin 36
epot 4
equ 26
uip 36
uit 9
ered 26
ep 3
ig 3
m 29
od 3
r p. 24
ug 4
utt 10
escap 36
cot 36
cud 36
ecr 10
fri 36
fum 36
ib 34
il 23
it 26
mer 36
mol 36
or 26
orn 10
pad 36
pej 36
pir 10
pit 36
tib. 36
tof 36
um 26
etern 10
evag 4
ers 24
ir p. 24
it 29
ol 27
olv 6
ulg 9
exalb 24
exalt 36
cav 9
cit 9
cud 6
cus 9
ecr 36
ed 3
ent 36
hal 9
hum 36
im 3
ist 6
jur 24
lid 6
opt 9
or p. 25
orn 36
oss 9
pet 6
pi 29
pil 9
pon 6
put 9
sec 24
ser 6
sol 9
sud 9
sug 6
ul p. 25
und 9
ur 3

F

fabe 29
bul 36
cet 24
d 27
j 29
lc 17
ls p. 26
mul 9
on 23
rt 29
tur 36
febru 9
ch 27
d 34
ltr 10
m 29
ri 27
rit 24
rm 23
rr 29
t 23
ud 27
fid 26
gul 9
gur 10
fij 27
li 29
ltr 36
n 34
nd 3
or 34
rm 29
sg 17
ss 23
ut 23
flagr 9
amm 9
ech 36
ect 11
et 29
ex 4
ig 3
uid 24
ut 4
focil 9
der 10
dic 24
ed p. 26
gl 34
gn 29
li 27
rb 34
rci 36
rj 29
rm p. 27
rz 26
s 29
ss p. 27
frang 11
anj 36
app 10
aud 9
eg 17
eil 36
em 27
en p. 28
ett 10
is 29
od 23
omb 10
ot 27
u 23
um 3
ust 10
ut 29
fucil 10
gg 34
git 9
lc 34
m 27
nd 27
ner 9
ng 3
ri p. 28
furt 23
sc p. 29
sil 36
tu 3

G

gabb 23
le p 29
lib 36
n 27
nz 26
rb 29
rz 23
st 27
y 29
gel p. 30
ner 9
rib 36
ss 26
st p. 30
giamb 10
b 29
el 25
gn 3
ov 25
tt 26
unt 10
ur 25
glaci 9
ob p. 31
os 25
ub 3
gol 25
m 27
rm 29
grab 27
adu 36
am 25
an 29
ass 10
att 10
av p. 31
az 34
eg p. 32
em 34
id 25
if 26
ond 10
u 29
upp 10
guad 25
ar 34
ard 36
azz 10
eld 36
id 26
in 27
guind 36
ip 29
st 29
gyps p. 32
r p. 33

H

habit 9
bl 29
cin 36
d 29
l p. 33
ll 27
rt 27
hebet 9
n 27
rb 29
red 36
hiem p. 34
lar 9
nc 17
honor 36
rad 36
hum p. 34
rt 29
sm 29

I

idu 4
ignav 24
nor 9
illim 9
lin 11
lud 11
lum 10
iman 29
bas 10
bib 11
bol 10
bu 3
it p. 35
mel 10
mit 10
mol 9
mor 10
mut 9
pal 10
pep 10
per 9
pet 11
pi p. 35
pon 11
pur 10
put 36

inaj 25
alb 24
am 26
aqu 14
aur 14
cal 24
car 14
cer 9
cib 24
co 29
coj 14
col 11
cub 36
cud 11
dag 9
dic 9
dor 14
du 25
duc 11
dur 9
err 24
esc 9
fam 9
fil 14
fl p. 36
for 24
fri 24
fun 14
gel 14
ger 11
hal 36
hi p. 36
hum 24
ib 34
ici 36
ig 3
iqu 24
mol 36
mut 36
nav 14
nod 9
nom 14
nov 36
occ 9
ond 14
or 26
orn 24
ram 14
sal 14
set 14
sip 11
sof 14
sol 9
st 29
sum 11
teg 11
ter 14
tim 9
ton 24
intr p. 37
tu 3
tuf 14
um 25
unc 9
und 36
ung 11
ur 3
vad 11
vas 14
ve 34
veh 11
vej 36
vit 9
voc 9
vol 14
iris 27
rad 11
rep 11
ret 14
rig 9
rit 36
ror 14
ru 3
ruf 9
iscus 14
ol 26
tri 36
iter p. 37

J

jabel 36
ch 29
ct p. 38
di 29
m 29
mbr 36
rop 36
jet 29
jorob 36
rr 27
jubil 24
dic 9
g p. 38
nel 36
nt 29
r 27
rgi 24
ss 3
st 4

L

labor 9
br 29
cer 14
cr 27
lact 29
cun 9
et p. 39
id 26
ld 34
mb 3
min 36
ni p. 39
nz 17
pid 9
qu 29
rd 25
rg 34
rv p. 40
ss p. 40
str 16
tin 16
tt 26
ud p. 41
vor 16
leg 3
gr 27
m 26
mbr 36
n 29
nt p. 41
nz 26
p 27
tam 16
tr 29
ud 29
vig 9
vit 16
libr 25
cu 29
di 27
ev 25
g p. 42
ll 29
m 30
mpi 36
ne p. 42
qu p. 43
si 30
st 27
tur 24
vid 24
vr 26
lod 25
gor 16
gr 30
ngi 7
rd 26
ric 9
s 27
tt 25
luch 30
cid 16
cr p. 43
ludi 30
min 9
str 9
t 26
x p. 44
lymph 9

M

macer 9
ci 4
cin 16
did 9
dur 36
est 7
gn 26
go 30
lax 9
lt 34
lus 16
lv 30
mm 25
n p. 44
nch 36
nd 25
nej 36
ns 26
ran 36
rid 36
rit 9
rn 25
rz 26
ss 25
tur 9
ul 30
yor 36
zn 30
med 30
ll 30
mor 9
nom 16
nor 36
rit 16
rl 25
rm 27
rs p. 45
rt 26
s 27
tu 3
mim 27
ngu 36
ni p. 45
nor 9
nu 34
r 4
rl 27
rr 25
sur 16
mitig 9
tr 26
tt 3
moch 27
del 36
der 9
dol 16
j 30
ll 27
ned 36
net 16
nt 30
str 16
t 4
tej 36
y 27
zz 26
mud 27
lt p. 46
n 34
nd 26
r 25
rci 36
sit 36
ss 4
til 36
tu 25

N

narr p. 46
s 26
sc 3
sp 25
tur 16
v p. 47
nec p. 47
ct 3
gum 7
rb 25
rvi 36
tt 25
niel 30
ell 16
gr p. 48
tid 7
nod p. 48
dr 34
j 26
min 16
t 25
vel 36
ver 16
nub 3
bl 27
d 25
mer 9
nunti 9
tr 34

O

obar 4
bib 11
duc 11
dur 9
ed 3
es 4
ger 11
hum 7
jet 36
lid 11
nub 11
rod 11
sip 7
su 3
teg 11
tur 36
unc 7
vel 7
vi 30
occid 11
cul 11
cup 12
ul 4
ult 36
up 30
odi 30
or 26
ofens 36
fig 11
foc 12
ici 36
ojal 30
olet 4
ezz 16
f 26
iv 31
tr 26
vid 36
ombr 26
in 4
ondul 36
er p. 49
est 16
or 25
t 34
ust 12
opac 1
er 31
ert 7
il 31
im 1
pet 11
pic 7
pon 11

opput 7
res 36
t 1
ugn 36
orden 36
din 12
f 27
ill 36
in 31
l 25
m 26
n 1
oscit 7
cur 16
pit 16
sid 16
ottur 16
ovatt 16
ic 4
oxid 31

P

pact 27
li 31
lm 27
lp 1
nd 1
ng 3
nn 26
rc 3
ri 1
rod 16
sm 31
tr 1
ur 34
ut 31
ves 16
vid 7
von 36
pect 31
d 1
in 31
jor 7
l 25
ll 3
nd 3
ns 26
or 31
rag 11
red 11
rfl 12
rn 25
ror 16
rur 11
sc 17
pest 26
t 3
pian 25
f 27
ll 25
nch 36
nor 36
ns 3
nz 25
omb 16
ot 25
p 25
s 4
st 31
um 25
plag 1
ant 12
asm 12
eg 17
ic 1
om 31
or 1
um 1
pol 34
mp 4
n 3
nd 25
nz 26
pol 16
pp 25
pul 12
rt 1
sc 3
t 26
praed 12
anz 16
eci 36
ed 25
em 3
ens 36
ess 7
est 16
ezz 16
iv 1
ob 1
ofl 12
ont 8
oru 11
ov 26
pugn 25
j 31
ls 1
mic 12
n 27
nt 25
ntu 36
rg 1
t 1

Q

quadr 12
aer 11
art 8
el 27
em 31
et 25
it 27
ot 26

R

racem 7
di 1
ff 25
mp 25
p 26
pt 1
s 25
sp 31
uc 4
un 25
reag 3
am 27
ar 31
cit 12
cog 11
col 7
cus 12
dam 7
dd 3
don 7
dr 27
fig 11
fl 1
fug 7
gal 8
gel 12
gl 31
gn 1
lax 12
leg 11
lid 11
lig 12
lu 34
me 1
n 25
nd 31
nt 27
nu 3
orn 7
pet 11
san 12
ser 12
st 31
sum 11
ret 4
tex 11
veh 11
vom 11
rialz 8
ar 25
att 8
cap 8
dam 8
don 8
for 8
gid 7
gir 8
gor 12
lim 8
lod 8
lut 8
men 8
mur 8
nom 8
orn 8
pi 31
pos 8
san 8
sit 7
un 34
vel 8
zz 25
rob 25
bl 27
ci 31
g 1
ll 31
tol 8
vin 8
rub 26
col 8
ct 1
gum 8
in 31
m 25
mp 3
nc 4
sp 4
til 12
tt 26

S

sacc 1
ch 27
cr 25
gin 12
gr 26
in 32
ld 26
salt 25
lut 12
lv 1
n 26
nc 34
por 18
rp 26
ti 1
uci 12
sbad 34
and 18
scab 3
af 25
arn 18
av 26
erp 18
ind 11
ip 25
isc 11
oj 26
or 25
rib 11
rut 7
ud 25
ulp 11
ult 18
ur 34
us 26
seb 15
cur 18
d 27
duc 11
gn 26
jug 7
min 18
nt 34
par 12
qu 3
ren 12
rp 3
rr 34
rv 1
st 25
voc 12
vr 25
sfug 34
sibil 7
ep 25
gn 27
lb 32
m 1
mpl 7
nu 1
rp 15
s 32
st 3
ti 32
tu 25
smel 25
ir 25
snod 25
soas 32
b 32
ci 1
d 26
gn 25
j 25
lf 25
lid 12
lv 3
mm 26
mni 12
nd 32
pl 32
por 12
pr 25
rs 25
st 26
spaj 25
alm 18
and 11
ar 26
arg 11
arl 18
azz 18
ect 12
ed 34
egn 18
ens 18
er 1
es 26
ic 1
in 25
iss 12
oli 12
ons 7
os 25
ul 25
um 1
urc 12
ut 26
stagn 12
amp 18
as 26
az 25
ell 7
end 18
ern 11
igm 7
ill 18
ip 1
ir 26
irp 18
iv 25
izz 19
orn 19

strem 19
rep 11
ru 34
uf 26
upr 12
urb 19
ut 25
subar 7
big 11
blu 11
bur 11
cid 7
d 1
ffl 12
fl 32
gil 12
lc 1
mm 15
on 25
pin 12
pp 15
zz 26

T

tabul 7
cit 19
tali 1
min 7
p 32
rd 1
rj 32
rp 25
ss 26
st 25
x 1
tech 32
mer 12
mpr 19
mpt 12
nt 25
nu 1
pid 7
rs 32
rti 12
rz 25
sor 19
st 27
tr 15
tild 27
nt 27
r 32
tol 19
zn 32
told 33
ler 12
m 33
nd 26
pp 26
rn 33
rp 1
rt 34
s 25
trab 33
act 12
ad 34
an 1
at 27
em 26
ep 33
in 26
ipl 7
it 25
ot 26
ov 25
uf 33
us 15
tuff 25
mb 33
mul 12
nic 7
turb 26
rr 27
tel 19

U

uber 1
ulcer 12
ul 1
uman 25
br 1
ect 12
ett 19
uncin 19
izz 19
urin 15
t 26
ustil 12
urp 12
util 25
uvi 33

V

vaci 33
cu 1
vall 1
lor 19
lu 33
lut 19
nt 25
por 12
r 25
st 1
vect 1
get 12
len 19
lit 7
nd 27
nen 12
nt 26
rs 26
ss 25
tu 15
x 1
vibr 1
ci 27
dim 19
et 25
gi 33
gil 12
gn 26
lt 27
viol 1
r 27
sc 15
ti 1
vocit 12
it 25
lg 1
lt 26
m 34
r 1
t 33
vuot 25

Z

zaf 27
ff 26
mp 26
nj 33
rp 33
zel 26
pp 25
zinn 25
zomb 25
n 25
zurr 33

adam	**abam**	impossible to **accept**, price too high
„	**abamus**	„ „ „ „ „ low
„	**abant**	against acceptance
„	**abare**	met with an **accident**
„	**abaris**	„ „ a serious accident
„	**abatur**	a serious accident occurred
„	**abimur**	did not meet with an accident
„	**abis**	owing to accident
„	**abitis**	no accident
„	**abo**	for our (my) **account**
„	**amini**	„ your account
„	**amur**	„ their (his) account
„	**andam**	„ joint account
„	**andi**	„ a/c and at risk of [best
„	**andior**	we (I) leave you to **act** for the
„	**andius**	{if cannot act as proposed we (I) leave you to act for the best
„	**andos**	wire **address**
„	**ans**	next address
„	**antem**	address telegrams and letters
„	**antis**	„ „ „ „ care
„	**antium**	**administration** [of
„	**anto**	more **advantageous**
„	**antur**	consider it more advantageous
„	**aremus**	if more advantageous
„	**arent**	no **advice** received
„	**arere**	do you consider it advisable?
„	**ares**	consider it advisable
„	**aretis**	it does not seem advisable
„	**aretur**	not **affected** to any material extent
„	**ari**	affected to a considerable extent
„	**arier**	wire whether **agree**
„	**at**	provided you agree
„	**atae**	we (I) do not agree
„	**atarum**	if you do not agree
„	**atior**	„ „ „ „ „ wire
„	**atius**	against **allowance**
„	**atos**	we (I) grant allowance asked for
„	**atote**	wire allowance
„	**atum**	half **amount**
„	**aturae**	full „
„	**aturi**	wire what amount
„	**aturos**	out of the amount
„	**averam**	**appeal** dismissed
„	**averit**	„ „ with costs
„	**avero**	„ granted
„	**avimus**	it **appears** that
„	**avisse**	it does not appear
„	**avisti**	file **application**
„	**avit**	filed „
„	**emus**	this **applies** to
„	**ent**	on **approval**
„	**es**	get approval of
„	**etis**	to be settled by **arbitration**
„	**etur**	arbitrators awarded
„	**or**	endeavour to **arrange** (with)

ador	**abam**	**arrived** too late
„	**abamus**	safely arrived to-day
„	**abant**	to arrive
„	**abare**	on arrival
„	**abaris**	all **articles**
„	**abatur**	impossible to **ascertain**
„	**abimur**	**assets** and liabilities
„	**abis**	quick assets
„	**abitis**	value of assets
„	**abo**	estimate value of assets
„	**amini**	available assets
„	**amur**	„ „ nil
„	**andam**	assets of uncertain value
„	**andi**	assets realized
„	**andior**	out of assets
„	**andius**	„ „ „ realized
„	**andos**	wire **assortment**
„	**ans**	send „
„	**antem**	call **attention** to
„	**antis**	at **auction**
„	**antium**	to be sold at auction
„	**anto**	sold at auction and realized
„	**antur**	next auction sales
„	**aremus**	wire whether **authorized**
„	**arent**	total **available**
„	**arere**	wire total available
„	**ares**	out of total available
„	**aretis**	if available
„	**aretur**	wire whether available
„	**ari**	at an **average**
„	**arier**	above the average
„	**at**	below „ „
„	**atae**	general average
„	**atarum**	particular average
„	**atior**	in double **bags**
„	**atius**	wire **balance**
„	**atos**	wire our (my) credit balance
„	**atote**	balance in our (my) favor
„	**atum**	„ „ your favor
„	**aturae**	out of balance
„	**aturi**	to arrive in **ballast** latest on
„	**aturos**	arrived in ballast
„	**averam**	adjudged **bankrupt**
„	**averit**	in **barrels**
„	**avero**	on same **basis**
„	**avimus**	„ usual „
„	**avisse**	„ new „
„	**avisti**	on the basis of
„	**avit**	to **be**
„	**emus**	not to be
„	**ent**	will be
„	**es**	„ not be
„	**etis**	„ probably be
„	**etur**	„ „ not be
„	**or**	

adun	**abam**	from **beginning** (of)
„	**abamus**	at the beginning
„	**abant**	we (I) **believe**
„	**abare**	„ „ do not believe
„	**abaris**	there is reason to believe
„	**abatur**	do you believe?
„	**abimur**	it is generally believed
„	**abis**	for the **benefit** of
„	**abitis**	for your benefit
„	**abo**	do your **best**
„	**amini**	shall do our (my) best
„	**amur**	best possible
„	**andam**	slightly **better**
„	**andi**	no better
„	**andior**	much better
„	**andius**	or better if possible
„	**andos**	for want of anything better
„	**ans**	collect **bill**
„	**antem**	collected bill
„	**antis**	Bill has passed
„	**antium**	Bill likely to be passed
„	**anto**	B/L
„	**antur**	send B/L
„	**aremus**	sent B/L
„	**arent**	against B/L
„	**arere**	B/L to order
„	**ares**	B/L made out in our (my) name
„	**aretis**	in **blank**
„	**aretur**	signed in blank
„	**ari**	en **bloc**
„	**arier**	**Blue** Book
„	**at**	on **board**
„	**atae**	all on board
„	**atarum**	„ „ „ saved
„	**atior**	„ „ „ lost
„	**atius**	taken on board
„	**atos**	Board of Directors
„	**atote**	1 boatload
„	**atum**	2 boatloads
„	**aturae**	in **bond** to
„	**aturi**	interim **bonus**
„	**aturos**	final bonus
„	**averam**	**book** value
„	**averit**	in the **bottom**
„	**avero**	bottom damaged
„	**avimus**	**bound** for
„	**avisse**	in **boxes**
„	**avisti**	„ double boxes
„	**avit**	any **brand**
„	**emus**	wire what brand
„	**ent**	superior brand
„	**es**	owing to violent **break** in
„	**etis**	„ „ **breakage**
„	**etur**	less **brokerage**
„	**or**	including brokerage

aequ	**abam**	1/8 % brokerage
„	**abamus**	1/4 % „
„	**abant**	3/8 % „
„	**abare**	1/2 % „
„	**abaris**	3/4 % „
„	**abatur**	7/8 % „
„	**abimur**	1 % „
„	**abis**	1 1/2 % „
„	**abitis**	in **bulk**
„	**abo**	on **bull a/c**
„	**amini**	on the part of bulls
„	**amur**	are very **busy**
„	**andam**	impossible to **buy** price too high
„	**andi**	„ „ „ prices „ „
„	**andior**	wire at what price you can buy
„	**andius**	„ „ „ „ „ recommend buying
„	**andos**	wire quantity and price you can buy
„	**ans**	wire quantity and price you recommend buying
„	**antem**	buy for our (my) a/c
„	**antis**	bought for your a/c
„	**antium**	to be bought
„	**anto**	wire whenever consider advisable to buy
„	**antur**	buy and ship by first steamer
„	**aremus**	buy on most favorable terms
„	**arent**	shall buy against you
„	**arere**	have bought against you
„	**ares**	owing to chance purchase
„	**aretis**	this is quite a chance purchase
„	**aretur**	are in treaty to buy
„	**ari**	we (I) suggest buying
„	**arier**	against purchase
„	**at**	„ purchases
„	**atae**	2nd hand purchase
„	**atarum**	can buy 2nd hand
„	**atior**	bought from
„	**atius**	there are **buyers**
„	**atos**	„ „ „ but no **sellers**
„	**atote**	„ „ sellers
„	**atum**	„ „ „ but no buyers
„	**aturae**	„ „ eager buyers
„	**aturi**	„ „ „ sellers
„	**aturos**	„ „ many buyers
„	**averam**	„ „ „ sellers
„	**averit**	„ „ no buyers
„	**avero**	„ „ no sellers
„	**avimus**	„ „ hardly any buyers
„	**avisse**	„ „ „ „ sellers
„	**avisti**	nominal, no buyers
„	**avit**	there are large buyers
„	**emus**	buyers are reserved
„	**ent**	sellers „ „
„	**es**	by **cable**
„	**etis**	cable interrupted
„	**etur**	to **call** at
„	**or**	at call

aer	**abam**	option to call
„	**abamus**	do all you **can**
„	**abant**	shall do all we (I) can
„	**abare**	to be **cancelled**
„	**abaris**	wire terms for cancelling
„	**abatur**	until cancelled
„	**abimur**	option to cancel
„	**abis**	try to get option to cancel
„	**abitis**	cancelling date
„	**abo**	„ „ . . . days later
„	**amini**	**capital** has been increased
„	**amur**	„ will be increased
„	**andam**	owing to insufficiency of ca-
„	**andi**	on capital a/c [pital
„	**andior**	wire amount of capital
„	**andius**	uncalled capital
„	**andos**	unemployed capital
„	**ans**	estimate capital at
„	**antem**	**captain** reports
„	**antis**	recommend greatest possible
„	**antium**	do not care for [**care**
„	**anto**	small **cargo**
„	**antur**	large „
„	**aremus**	smaller „
„	**arent**	larger „
„	**arere**	1 full cargo
„	**ares**	2 „ cargoes
„	**aretis**	part of cargo
„	**aretur**	general cargo
„	**ari**	**carriage** paid
„	**arier**	in any **case**
„	**at**	in no „
„	**atae**	in that „
„	**atarum**	in case of
„	**atior**	**cash** at call
„	**atius**	payable in cash
„	**atos**	cash available
„	**atote**	out of cash available
„	**atum**	cash on hand not sufficient
„	**aturae**	out of amount cashed
„	**aturi**	remit amount cashed
„	**aturos**	remit when cashed
„	**averam**	spare cash
„	**averit**	no spare cash
„	**avero**	send **catalogue**
„	**avimus**	wire **cause** of
„	**avisse**	almost **certain**
„	**avisti**	**certificate**
„	**avit**	certificates
„	**emus**	send certificate
„	**ent**	sent certificate
„	**es**	send certificates
„	**etis**	legalized certificate
„	**etur**	„ certificates
„	**or**	send legalized certificate

affl	**abam**	**Chamber** of Commerce
„	**abamus**	there is no **chance**
„	**abant**	„ „ hardly any chance
„	**abare**	there is some chance
„	**abaris**	this is quite a chance
„	**abatur**	chances are against
„	**abimur**	„ „ in favor of
„	**abis**	wire whether there is any
„	**abitis**	if there is no chance [chance
„	**abo**	no **change** from last report
„	**amini**	wire **charges**
„	**amur**	free of charge
„	**andam**	inclusive of all charges
„	**andi**	to take charge of
„	**andior**	**charter** party
„	**andius**	as per charter party
„	**andos**	charterers' agent/s
„	**ans**	at charterers' risk and expense
„	**antem**	„ „ option
„	**antis**	send **cheque** for
„	**antium**	sent „ „
„	**anto**	crossed cheque
„	**antur**	send crossed cheque for
„	**aremus**	on cheque a/c
„	**arent**	wire balance cheque a/c
„	**arere**	cheque on London
„	**ares**	„ „ Paris
„	**aretis**	„ „ New-York
„	**aretur**	„ „ Berlin
„	**ari**	pay cheque
„	**arier**	do not pay cheque
„	**at**	have paid cheque
„	**atae**	cheque No.
„	**atarum**	in **chests** (of)
„	**atior**	half chests
„	**atius**	**choice** lot
„	**atos**	owing to **cholera**
„	**atote**	as per **circular**
„	**atum**	in **circulation**
„	**aturae**	according to **circumstances**
„	**aturi**	in full discharge of **claims**
„	**aturos**	claim for compensation
„	**averam**	counter-claim
„	**averit**	claim settled
„	**avero**	claim inadmissible
„	**avimus**	claim recognized
„	**avisse**	1st **class**
„	**avisti**	2nd class
„	**avit**	3rd class
„	**emus**	high class
„	**ent**	intermediate class
„	**es**	in order to **clear**
„	**etis**	for want of **coal**
„	**etur**	send 1 copy of this **code**
„	**or**	„ 2 copies „ ,

agit abam	in future the following cipher will mean the preceding 2 words
„ **abamus**	in future the following cipher will mean the preceding 3 words
„ **abant**	send for **collection**
„ **abare**	Bill sent for collection
„ **abaris**	**a collision** took place between
„ **abatur**	owing to collision
„ **abimur**	**commission** & guarantee
„ **abis**	against commission (of)
„ **abitis**	wire what commission
„ **abo**	„ „ „ and guarantee
„ **amini**	no commission
„ **amur**	inclusive of commission
„ **andam**	exclusive „ „
„ **andi**	1/4% commission
„ **andior**	1/2% „
„ **andius**	3/4% „
„ **andos**	1% „
„ **ans**	1 1/2% „
„ **antem**	2% „
„ **antis**	2 1/2% „
„ **antium**	3% „
„ **anto**	4% „
„ **antur**	5% „
„ **aremus**	7 1/2% „
„ **arent**	10% „
„ **arere**	reduced commission
„ **ares**	full commission
„ **aretis**	half „
„ **aretur**	**communicate** with
„ **ari**	communication interrupted
„ **arier**	telegraphic communication interrupted
„ **at**	communication restored
„ **atae**	**comparatively**
„ **atarum**	**compensation**
„ **atior**	against a compensation
„ **atius**	to be able to **compete**
„ **atos**	competition is very keen
„ **atote**	owing to **complaints** (about)
„ **atum**	not yet **complete**
„ **aturae**	to be completed
„ **aturi**	completed shipment
„ **aturos**	in **compliance** with
„ **averam**	owing to non-compliance with
„ **averit**	**compliments** of the season
„ **avero**	**composition** of .... in the £
„ **avimus**	„ „ 3/- „ „ „
„ **avisse**	„ „ 4/- „ „ „
„ **avisti**	„ „ 5/- „ „ „
„ **avit**	„ „ 6/- „ „ „
„ **emus**	„ „ 7/- „ „ „
„ **ent**	„ „ 8/- „ „ „
„ **es**	„ „ 9/- „ „ „
„ **etis**	„ „ 10/- „ „ „
„ **etur**	„ „ 11/- „ „ „
„ **or**	„ „ 12/- „ „ „

alt abam	**composition** of 13/- in the £
„ **abamus**	„ „ 14/- „ „ „
„ **abant**	„ „ 15/- „ „ „
„ **abare**	„ „ 16/- „ „ „
„ **abaris**	wire **condition** of
„ **abatur**	in good condition
„ **abimur**	in excellent condition
„ **abis**	in an exhausted condition
„ **abitis**	in a bad condition
„ **abo**	on condition that
„ **amini**	convey **condolences** (to)
„ **amur**	sincerest condolences
„ **andam**	deserving full **confidence**
„ **andi**	„ no „
„ **andior**	**confidential(ly)**
„ **andius**	this is strictly confidential
„ **andos**	confidently expected
„ **ans**	wire **confirmation**
„ **antem**	on confirmation
„ **antis**	sincere **congratulations**
„ **antium**	subject to **consent** of
„ **anto**	„ „ our (my) consent
„ **antur**	„ „ your consent
„ **aremus**	it is **considered** (that)
„ **arent**	„ „ generally considered
„ **arere**	provided you consider
„ **ares**	matter is under consideration
„ **aretis**	**consigned** to
„ **aretur**	wire name of consignee
„ **ari**	consignment
„ **arier**	consignments
„ **at**	on consignment a/c
„ **atae**	out of consignment
„ **atarum**	**consisting** principally of
„ **atior**	American **Consul**
„ **atius**	British Consul
„ **atos**	French Consul
„ **atote**	German Consul
„ **atum**	American Consulate
„ **aturae**	British Consulate
„ **aturi**	for **consumption**
„ **aturos**	for local consumption
„ **averam**	wire **contango** rate
„ **averit**	wire **contents**
„ **avero**	on the **Continent**
„ **avimus**	continuation rate
„ **avisse**	continuation rates
„ **avisti**	continue on same scale
„ **avit**	against **contract**
„ **emus**	„ contracts
„ **ent**	send contract
„ **es**	signed contract
„ **etis**	as per contract
„ **etur**	contract renewed until
„ **or**	contract expiring

amyl	**abam**	new contract
„	**abamus**	old contract
„	**abant**	owing to breach of contract
„	**abare**	completed all contracts
„	**abaris**	„ contract
„	**abatur**	under **control**
„	**abimur**	at your earliest **convenience**
„	**abis**	**cooperative** company
„	**abitis**	send **copy**
„	**abo**	send copies [dence
„	**amini**	send copy of all correspon-
„	**amur**	send copy of invoice
„	**andam**	**correct** as follows
„	**andi**	wire **cost**
„	**andior**	wire estimate of cost
„	**andius**	free of cost
„	**andos**	cost to be defrayed by
„	**ans**	would cost about
„	**antem**	reduced cost
„	**antis**	payment of **coupon**
„	**antium**	settlement by **Court**
„	**anto**	Prize Court
„	**antur**	against **credit**
„	**aremus**	renew credit
„	**arent**	Letter of Credit
„	**arere**	against Letter of Credit
„	**ares**	credit exhausted, renew
„	**aretis**	blank credit
„	**aretur**	to your credit
„	**ari**	to our (my) credit
„	**arier**	send Letter of Credit
„	**at**	no credit
„	**atac**	overdrawn credit
„	**atarum**	avoid exceeding credit
„	**atior**	wire to what extent credit might be granted
„	**atius**	limited credit
„	**atos**	enjoying unlimited credit
„	**atote**	out of credit
„	**atum**	moderate credit is looked upon as an ordinary risk
„	**aturae**	deserving no credit
„	**aturi**	telegraphic credit
„	**aturos**	on behalf of creditors
„	**averam**	meeting of creditors
„	**averit**	**crew** and passengers
„	**avero**	**crisis** is imminent
„	**avimus**	crisis has passed
„	**avisse**	old **crop**
„	**avisti**	new crop
„	**avit**	old and new crop
„	**emus**	out of old crop
„	**ent**	„ „ new „
„	**es**	wire estimate of crop
„	**etis**	crop estimated at
„	**etur**	we (I) estimate crop at
„	**or**	total outturn of crop

ancl	**abam**	world's crop
„	**abamus**	crop is a serious failure
„	**abant**	crop has begun to come in
„	**abare**	no damage to crop
„	**abaris**	it is too early to express an opinion on the crop
„	**abatur**	the area sown is
„	**abimur**	the crops look well
„	**abis**	wire estimate of exportable
„	**abitis**	at **Custom House** [surplus
„	**abo**	wire is **cut**
„	**amini**	owing to a **cyclone**
„	**amur**	badly **damaged**
„	**andam**	slightly damaged
„	**andi**	damage estimated at
„	**andior**	we (I) estimate damage at
„	**andius**	cargo damaged
„	**andos**	no damage to
„	**ans**	vessel damaged
„	**antem**	damages assessed at
„	**antis**	in serious **danger**
„	**antium**	no danger
„	**anto**	there is no danger at present
„	**antur**	„ appears to be no danger
„	**aremus**	out of danger
„	**arent**	wire whether there is any
„	**arere**	wire **date** of [danger
„	**ares**	to be dated
„	**aretis**	after date
„	**aretur**	up to date
„	**ari**	B/L dated
„	**arier**	final date
„	**at**	30 days after date
„	**atac**	60 „ „ „
„	**atarum**	90 „ „ „
„	**atior**	date of departure
„	**atius**	„ „ arrival
„	**atos**	wire date of departure
„	**atote**	„ „ „ arrival
„	**atum**	at **dawn**
„	**aturae**	at dawn this morning
„	**aturi**	all **day**
„	**aturos**	day and night
„	**averam**	during last few days
„	**averit**	in a day or two
„	**avero**	to-day
„	**avimus**	to-day's
„	**avisse**	on same day
„	**avisti**	every other day
„	**avit**	working days
„	**emus**	**deadweight** not more than
„	**ent**	balance deadweight
„	**es**	too **dear**
„	**etis**	**death** reported of
„	**etur**	wire **decision**
„	**or**	on **deck**

anim	**abam**	deck cargo	
„	**abamus**	to be **declared**	
„	**abant**	send **deed** of	
„	**abare**	owing to **deficit**	
„	**abaris**	without **delay**	
„	**abatur**	will be delayed until	
„	**abimur**	to be **delivered**	
„	**abis**	delivery 1 week	from receipt of order
„	**abitis**	„ 2 weeks	
„	**abo**	„ 3 „	
„	**amini**	„ 4 „	
„	**amur**	„ 5 „	
„	**andam**	„ 6 „	
„	**andi**	„ 2 months	
„	**andior**	„ 3 „	
„	**andius**	delivery as soon as possible	
„	**andos**	late delivery	
„	**ans**	owing to late delivery	
„	**antem**	if goods delivered too late, will buy against you	
„	**antis**	goods delivered too late, have bought against you	
„	**antium**	if delivered too late	
„	**anto**	cannot deliver in time	
„	**antur**	wire shortest possible delay of delivery	
„	**aremus**	for forward delivery	
„	**arent**	consumptive **demand**	
„	**arere**	stocks sufficient for demand	
„	**ares**	to pay **demurrage**	
„	**aretis**	on **departure**	
„	**aretur**	departure delayed	
„	**ari**	on **deposit**	
„	**arier**	to be deposited	
„	**at**	against deposit (of)	
„	**atae**	out of deposit	
„	**atarum**	deposit receipt	
„	**atior**	„ receipts	
„	**atius**	„ rate	
„	**atos**	„ rates	
„	**atote**	wire deposit rate	
„	**atum**	require a deposit of	
„	**aturae**	all **descriptions**	
„	**aturi**	leaving much to be **desired**	
„	**aturos**	wire **destination**	
„	**averam**	at „	
„	**averit**	destination unknown	
„	**avero**	completely **destroyed**	
„	**avimus**	full **details** by this mail's letter	
„	**avisse**	details are lacking	
„	**avisti**	goods **detained**	
„	**avit**	to the **detriment** of	
„	**emus**	**development**	
„	**ent**	wire **diameter**	
„	**es**	**diamond** mines	
„	**etis**	**died** of cholera	
„	**etur**	died of plague	
„	**or**	died from fever	

ann	**abam**	died from sunstroke
„	**abamus**	owing to **difference** (in)
„	**abant**	owing to differences
„	**abare**	very **difficult**
„	**abaris**	owing to difficulties
„	**abatur**	**disbursements**
„	**abimur**	wire disbursements
„	**abis**	vessel ready to **discharge**
„	**abitis**	wire rate of **discount**
„	**abo**	1% discount
„	**amini**	1½% „
„	**amur**	2% „
„	**andam**	2½% „
„	**andi**	3% „
„	**andior**	3½% „
„	**andius**	4% „
„	**andos**	4½% „
„	**ans**	5% „
„	**antem**	6% „
„	**antis**	Bank of England rate of discount
„	**antium**	discount bill
„	**anto**	discount bills
„	**antur**	use your **discretion**
„	**aremus**	**dismissed** with costs
„	**arent**	at their (his) **disposal**
„	**arere**	„ disposal (of)
„	**ares**	„ your disposal
„	**aretis**	„ our (my) disposal
„	**aretur**	goods are at your disposal
„	**ari**	in **dispute**
„	**arier**	dispute settled
„	**at**	in **distress**
„	**atae**	**dividend** payment
„	**atarum**	interim dividend
„	**atior**	3% dividend
„	**atius**	3½% „
„	**atos**	4% „
„	**atote**	4½% „
„	**atum**	5% „
„	**aturae**	6% „
„	**aturi**	7% „
„	**aturos**	7½% „
„	**averam**	8% „
„	**averit**	10% „
„	**avero**	if cannot **do as asked,** wire what you can do
„	**avimus**	do your utmost
„	**avisse**	shall do our (my) utmost
„	**avisti**	in **dock**
„	**avit**	send **document**
„	**emus**	sent „
„	**ent**	send documents
„	**es**	sent documents
„	**etis**	**domiciliated**
„	**etur**	no **doubt**
„	**or**	very doubtful

arm	**abam**	at due date of **draft**
„	**abamus**	wire due date of draft(s)
„	**abant**	draft sent for collection
„	**abare**	have draft accepted
„	**abaris**	return draft accepted
„	**abatur**	wire whether you returned draft accepted
„	**abimur**	returned draft accepted
„	**abis**	returning draft accepted as soon as goods are found in order
„	**abitis**	draft accepted
„	**abo**	draft not accepted
„	**amini**	draft not advised
„	**amur**	pay draft
„	**andam**	pay drafts
„	**andi**	draft has been paid
„	**andior**	„ „ not been paid
„	**andius**	cancel draft
„	**andos**	draft payable
„	**ans**	drafts payable
„	**antem**	complete draft
„	**antis**	provisional draft
„	**antium**	partly **due** to
„	**anto**	probably due to
„	**antur**	from due date
„	**aremus**	wire due date
„	**arent**	at due date
„	**arere**	send **duplicate**
„	**ares**	sent „
„	**aretis**	send duplicates
„	**aretur**	free of **duty**
„	**ari**	new duty
„	**arier**	old duty
„	**at**	exclusive of duty
„	**atae**	inclusive of duty
„	**atarum**	duty increased (to)
„	**atior**	duty reduced „ [ed (to)
„	**atius**	duty will probably be increased (to)
„	**atos**	duty will probably be reduced (to)
„	**atote**	new duty will be in force on and after
„	**atum**	new duty applicable only on goods dispatched after
„	**aturae**	wire rate of duty
„	**aturi**	ad valorem duty
„	**aturos**	on basis of present duty
„	**averam**	it is too **early**
„	**averit**	as early as possible
„	**avero**	at the very earliest
„	**avimus**	**earnest** money
„	**avisse**	net **earnings**
„	**avisti**	1st **edition**
„	**avit**	2nd „
„	**emus**	new „
„	**ent**	to that **effect**
„	**es**	to be effected
„	**etis**	effect will only be temporary
„	**etur**	**either** or both
„	**or**	either of them

avoc	**abam**	**embargo** placed on vessel
„	**abamus**	„ „ „ goods
„	**abant**	„ removed
„	**abare**	owing to **embezzlement** of funds
„	**abaris**	in case of **emergency**
„	**abatur**	are (is) well **employed**
„	**abimur**	to **enable** us (me) to
„	**abis**	at the **end** of
„	**abitis**	almost at an end
„	**abo**	will end on
„	**amini**	are (am) **endeavouring** to
„	**amur**	shall endeavour as far as possible to
„	**andam**	**endorsed** in blank
„	**andi**	for want of endorsement
„	**andior**	**engagements**
„	**andius**	fulfilled outstanding engagements
„	**andos**	engagements are heavy
„	**ans**	**engines** disabled
„	**antem**	not **enough** (for)
„	**antis**	if not enough
„	**antium**	**epidemic** broken out
„	**anto**	owing to epidemic
„	**antur**	epidemic disappeared
„	**aremus**	at the **equivalent** of
„	**arent**	or the equivalent
„	**arere**	owing to **error**
„	**ares**	not **essential**
„	**aretis**	**established** since
„	**aretur**	establishment
„	**ari**	establishments
„	**arier**	wire **estimate**
„	**at**	estimated at
„	**atae**	we (I) estimate (at)
„	**atarum**	wire market estimate
„	**atior**	confirm previous estimate
„	**atius**	official estimate
„	**atos**	above the estimate
„	**atote**	below the estimate
„	**atum**	at all **events**
„	**aturae**	in the event of
„	**aturi**	**ever** since
„	**aturos**	almost **every**
„	**averam**	to give **evidence**
„	**averit**	further evidence
„	**avero**	confirmatory evidence
„	**avimus**	it is evident that
„	**avisse**	**ex steamer**
„	**avisti**	**examine** again
„	**avit**	shall examine and wire
„	**emus**	examined and found correct
„	**ent**	without any **exception**
„	**es**	exceptional(ly)
„	**etis**	wire rate of **exchange**
„	**etur**	„ rates „ „
„	**or**	exchange is fluctuating

deck– p. 6

basi	abam	secure exchange
„	abamus	secured „
„	abant	calculate exchange at
„	abare	calculating exchange at
„	abaris	Stock Exchange
„	abatur	**exempt** from
„	abimur	to be exempt from
„	abis	nearly **exhausted**
„	abitis	exhausted, replace (renew)
„	abo	do you **expect**?
„	amini	we (I) do not expect
„	amur	when do you expect?
„	andam	provided you expect
„	andi	unless you expect
„	andior	it is generally expected that
„	andius	not expected
„	andos	daily expected
„	ans	whenever **expedient**
„	antem	at **expense** and risk of
„	antis	at our expense
„	antium	at your expense
„	anto	at buyer's expense
„	antur	at seller's expense
„	aremus	at ship's expense
„	arent	paying all expenses
„	arere	by **expert**
„	ares	by experts
„	aretis	expert reports
„	aretur	to be examined by experts
„	ari	wire **explanation**
„	arier	owing to **explosion**
„	at	for **export**
„	atac	try to get **extension** of time
„	atarum	grant extension of time
„	atior	wire to what **extent**
„	atius	to some extent
„	atos	to that extent
„	atote	to the extent desired
„	atum	**extra**ordinary
„	aturae	**face** value
„	aturi	is this a **fact**?
„	aturos	if this is a fact
„	averam	**failed** once
„	averit	„ twice
„	avero	„ several times
„	avimus	„ liabilities amount to
„	avisse	„ engagements are heavy
„	avisti	„ „ are unimportant
„	avit	„ owing to vicious speculation
„	emus	„ owing to failure of
„	ent	„ „ „ insufficiency of capital
„	es	failure is imminent
„	etis	completely failed
„	etur	involved in failure
„	or	not involved in failure

cac	abam	not involved in failure to any material extent
„	abamus	involved in failure to some extent
„	abant	„ „ „ „ a large „
„	abare	„ „ „ to a very large extent
„	abaris	wire whether involved in failure
„	abatur	single **fare**
„	abimur	double fare
„	abis	intermediate fare
„	abitis	1st class fare
„	abo	2nd class fare
„	amini	in your **favor**
„	amur	in our (my) favor
„	andam	in their (his) favor
„	andi	more favorable
„	andior	most „
„	andius	as favorable as possible
„	andos	an uneasy **feeling** prevails
„	ans	an easier „ „
„	antem	free from **fever**
„	antis	yellow fever
„	antium	yellow fever on board
„	anto	there are **few**
„	antur	very few
„	aremus	there are (is) very few
„	arent	**filling** up with
„	arere	**financial** position
„	ares	„ position is favorable
„	aretis	„ „ is excellent
„	aretur	„ „ „ very sound
„	ari	„ „ „ fairly good
„	arier	„ „ „ better now
„	at	„ „ „ unfavorable
„	atac	„ „ „ not sound
„	atarum	„ „ „ not quite sound
„	atior	„ „ „ highly unfavorable
„	atius	„ „ „ doubtful
„	atos	„ „ „ inascertainable
„	atote	„ „ „ critical
„	atum	wire re financial position
„	aturae	to pay **fine**
„	aturi	expect to **finish**
„	aturos	not yet finished
„	averam	by **fire**
„	averit	on fire
„	avero	fire broken out
„	avimus	destroyed by fire
„	avisse	fire not yet extinguished
„	avisti	fire extinguished
„	avit	„ „ damage
„	emus	„ „ „ estimated at
„	ent	well-known **firm**
„	es	firm unknown
„	etis	firm amalgamated with
„	etur	will retire from firm
„	or	retired from firm

cael	**abam**	from **first** to last
„	**abamus**	not **fit** for
„	**abant**	not fit for the purpose
„	**abare**	to be **fixed**
„	**abaris**	definitely fixed
„	**abatur**	yellow **flag**
„	**abimur**	white flag
„	**abis**	**floating** debt
„	**abitis**	during **fog**
„	**abo**	all **following**
„	**amini**	the following is (are)
„	**amur**	will follow shortly
„	**andam**	per **foot** run
„	**andi**	per foot super
„	**andior**	in **force**
„	**andius**	to come into force
„	**andos**	in force until
„	**ans**	in force until cancelled
„	**antem**	new **form**
„	**antis**	old form
„	**antium**	every **fortnight**
„	**anto**	in a fortnight
„	**antur**	cannot be found
„	**aremus**	**free** from
„	**arent**	guaranteed free from
„	**arere**	f. o. r. t.
„	**ares**	more freely
„	**aretis**	less freely
„	**aretur**	wire **freight**
„	**ari**	freight and charges
„	**arier**	chance freight
„	**at**	owing to chance freight
„	**atae**	can probably get chance freight
„	**atarum**	through freight
„	**atior**	deadweight freight
„	**atius**	homeward freight
„	**atos**	outward freight
„	**atote**	on the **frontier**
„	**atum**	owing to non-**fulfilment** (of)
„	**aturae**	in **full**
„	**aturi**	fully up to
„	**aturos**	are short of **funds**
„	**averam**	necessary funds
„	**averit**	unrequired funds
„	**avero**	do you require funds?
„	**avimus**	provide funds for
„	**avisse**	funds are sufficient (for)
„	**avisti**	no **further**
„	**avit**	in **future**
„	**emus**	future is yet uncertain
„	**ent**	in the immediate future
„	**es**	owing to a **gale**
„	**etis**	now **going** on
„	**etur**	holding **good** for
„	**or**	without **guarantee**

cant	**abam**	guaranteeing
„	**abamus**	for our (my) **guidance**
„	**abant**	for your guidance
„	**abare**	pleaded **guilty**
„	**abaris**	**1st hand**
„	**abatur**	2nd hand
„	**abimur**	in our (my) hands
„	**abis**	in your hands
„	**abitis**	not to hand
„	**abo**	must have a free hand
„	**amini**	to be handed over to
„	**amur**	owing to rough **handling**
„	**andam**	nothing unusual **happened**
„	**andi**	in **harbour**
„	**andior**	Refuge Harbour
„	**andius**	there is **hardly** any
„	**andos**	clean bill of **health**
„	**ans**	clean bills of health
„	**antem**	foul bill of health
„	**antis**	state of health
„	**antium**	wire state of health
„	**anto**	state of health is good
„	**antur**	state of health has improved
„	**aremus**	state of health is bad
„	**arent**	state of health is critical
„	**arere**	cargo **heated**
„	**ares**	very **heavy**
„	**aretis**	too heavy
„	**aretur**	will be **here**
„	**ari**	it is deemed preferable to **hold** aloof
„	**arier**	hold aloof
„	**at**	holders asking
„	**atae**	holder of the Bill
„	**atarum**	to-day is a **holiday** here
„	**atior**	to-morrow will be holiday here
„	**atius**	owing to the holidays
„	**atos**	send **home**
„	**atote**	sent home
„	**atum**	to be sent home
„	**aturae**	there is no **hope**
„	**aturi**	wire whether there is any hope
„	**aturos**	more hopeful feeling prevailing
„	**averam**	**horse**-power
„	**averit**	in **Hospital**
„	**avero**	**House** of Commons
„	**avimus**	wire **how**
„	**avisse**	wire how many
„	**avisti**	wire how much
„	**avit**	how are you?
„	**emus**	owing to **hurri**cane
„	**ent**	are (am) in a hurry (to)
„	**es**	do not be in a hurry (to)
„	**etis**	**hypo**thecary
„	**etur**	hypothecated
„	**or**	

cit abam	**identical** with
„ abamus	identified himself as
„ abant	**if** so
„ abare	if not
„ abaris	„ at all
„ abatur	„ it
„ abimur	„ you
„ abis	„ we (I)
„ abitis	„ they
„ abo	even if
„ amini	owing to **illness**
„ amur	**immediately**
„ andam	immediately if possible
„ andi	**immovable** property
„ andior	no immovable property
„ andius	owning immovable property
„ andos	for **import**
„ ans	not **important**
„ antem	very important
„ antis	this is very important
„ antium	it is **impossible** to
„ anto	if impossible to
„ antur	**impression** prevails (that)
„ aremus	very **improbable**
„ arent	greatly **improved**
„ arere	owing to improvements
„ ares	owing to **inade**quacy (of)
„ aretis	utterly **incap**able
„ aretur	not **included**
„ ari	if inclusive of
„ arier	**income** tax
„ at	**incomp**rehensible
„ atae	**incorp**orated
„ atarum	to be **increased**
„ atior	increasing gradually
„ atius	increasing largely
„ atos	will be considerably increased
„ atote	for an **indefinite** period
„ atum	**indemnifying**
„ aturae	**inde**pendent(ly) of
„ aturi	**indis**pensable
„ aturos	consider indispensable
„ averam	**inex**pedient
„ averit	considered inexpedient
„ avero	use your **influence**
„ avimus	we are (I am) **informed** (that)
„ avisse	wire information re
„ avisti	file information
„ avit	filed „
„ emus	received no information
„ ent	cannot give any information
„ es	according to trustworthy information
„ etis	no trustworthy information obtainable
„ etur	send full information by letter
„ or	wire information and send full details by letter

clar abam	information received is satisfactory
„ abamus	information received is very satisfactory
„ abant	information received is unsatisfactory
„ abare	**infring**ement
„ abaris	**injunct**ion against
„ abatur	interim injunction
„ abimur	badly **injured**
„ abis	slightly injured
„ abitis	not injured
„ abo	shall **inquire** and wire later on
„ amini	**insist** upon
„ amur	**insolvency** reported
„ andam	owing to insolvency
„ andi	by **instalments**
„ andior	at the **instance** of
„ andius	**instead** of
„ andos	wire **instructions**
„ ans	follow instructions (of)
„ antem	until further instructions
„ antis	**insufficient** (for)
„ antium	if insufficient
„ anto	to be **insured** (for)
„ antur	policy of insurance
„ aremus	send policy of insurance
„ arent	inclusive of insurance
„ arere	covered by insurance
„ ares	provisional insurance
„ aretis	**intended** for
„ aretur	as intended
„ ari	intentional(ly)
„ arier	wire rate of **interest**
„ at	annual interest
„ atae	pay interest
„ atarum	paying interest
„ atior	paid interest
„ atius	interest due
„ atos	with interest from
„ atote	interest on
„ atum	2 1/2 % annual interest
„ aturae	3 % „ „
„ aturi	3 1/2 % „ „
„ aturos	4 % „ „
„ averam	4 1/2 % „ „
„ averit	5 % „ „
„ avero	5 1/2 % „ „
„ avimus	6 % „ „
„ avisse	7 % „ „
„ avisti	prejudicial to your interests
„ avit	„ „ our interests
„ emus	**interfere** on behalf of
„ ent	from the **interior**
„ es	in the interior
„ etis	**inter**mediary
„ etur	intermediate
„ or	**inter**ruption

coen	**abam**	at an **interval** of
„	**abamus**	**intervention**
„	**abant**	for **invest**ment
„	**abare**	at the **invit**ation of
„	**abaris**	send **Invoice**
„	**abatur**	send copy of Invoice
„	**abimur**	proforma Invoice
„	**abis**	send proforma Invoice
„	**abitis**	as per Invoice
„	**abo**	Invoice amount
„	**amini**	not **involved**
„	**amur**	wire whether involved
„	**andam**	**irrespective** of
„	**andi**	not yet **issued**
„	**andior**	at issue
„	**andius**	**jettisoned** part cargo
„	**andos**	as far as can be **judged** (from)
„	**ans**	Court given judgment against
„	**antem**	judgement postponed
„	**antis**	all **kinds** of
„	**antium**	**know** nothing of
„	**anto**	do not know
„	**antur**	not known here [here
„	**aremus**	firm (person) named not known
„	**arent**	nothing positive as yet known
„	**arere**	with **label**
„	**ares**	with our label
„	**aretis**	**labour** is scarce
„	**aretur**	labour is cheap
„	**ari**	owing to **lack** of
„	**arier**	**landed** and warehoused
„	**at**	fresh landed
„	**atae**	abnormally **large**
„	**atarum**	will **last** until
„	**atior**	**late** of
„	**atius**	not **later** than
„	**atos**	too late
„	**atote**	if too late
„	**atum**	if too late wire
„	**aturae**	late shipped
„	**aturi**	will become **law**
„	**aturos**	wire name of lawyer
„	**averam**	**laydays** to commence
„	**averit**	„ „ „ not after
„	**avero**	owing to **leakage**
„	**avimus**	owing to heavy leakage
„	**avisse**	at **least**
„	**avisti**	on **leave**
„	**avit**	on sick leave
„	**emus**	left to-day
„	**ent**	left yesterday
„	**es**	left here to-day
„	**etis**	left here yesterday
„	**etur**	will leave to-day
„	**or**	will leave to-morrow

coir	**abam**	ready to leave
„	**abamus**	to be ready to leave
„	**abant**	shall take **legal** steps
„	**abare**	if not, shall take legal steps
„	**abaris**	avoid legal steps
„	**abatur**	do you recommend legal
„	**abimur**	**less** than [steps?
„	**abis**	not less than
„	**abitis**	if not less (than)
„	**abo**	**last letter** received dated
„	**amini**	on receipt of letter
„	**amur**	full particulars by letter
„	**andam**	„ „ „ to-day's
„	**andi**	wait our (my) letter (of) [letter
„	**andior**	**liabilities**
„	**andius**	net liability
„	**andos**	gross liability
„	**ans**	ranking liabilities
„	**antem**	outstanding liabilities
„	**antis**	liabilities secured
„	**antium**	liabilities are all unsecured
„	**anto**	liabilities returned at
„	**antur**	wire estimate of liabilities
„	**aremus**	available liabilities
„	**arent**	„ „ nil
„	**arere**	liable for
„	**ares**	not liable for
„	**aretis**	to be held liable (for)
„	**aretur**	you are at **liberty** to
„	**ari**	to be at liberty to
„	**arier**	first **lien** (on)
„	**at**	no **lives** lost
„	**atae**	too **light**
„	**atarum**	into lighters
„	**atior**	ex lighter
„	**atius**	not **likely** (to)
„	**atos**	wire whether likely (to)
„	**atote**	most likely
„	**atum**	wire **limit**
„	**aturae**	„ limits
„	**aturi**	avoid exceeding limit
„	**aturos**	„ „ limits
„	**averam**	above limit
„	**averit**	below limit
„	**avero**	limit is too high
„	**avimus**	„ „ „ low
„	**avisse**	limits are too high
„	**avisti**	„ „ „ low [low
„	**avit**	impossible to buy, limit is too
„	**emus**	impossible to sell, limit is too high
„	**ent**	limit in force for to-day
„	**es**	„ „ „ until
„	**etis**	„ „ „ „ cancelled
„	**etur**	limit in force until to-morrow evening
„	**or**	

crax	abam	at your limit
„	abamus	„ „ limits
„	abant	„ our limit
„	abare	„ „ limits
„	abaris	cancel outstanding limit
„	abatur	„ all outstanding limits
„	abimur	by **liner**
„	abis	in **liquidation**
„	abitis	will go into liquidation
„	abo	send **list** (of)
„	amini	send full list (of)
„	amur	list opens
„	andam	list will close
„	andi	very **little**
„	andior	too little
„	andius	at **Lloyd's**
„	andos	classed at Lloyd's
„	ans	will **load**
„	antem	to load
„	antis	vessel ready to load
„	antium	prompt loading
„	anto	early loading
„	antur	call **loans**
„	aremus	short loans
„	arent	loan rate
„	arere	raising loan
„	ares	loan floated
„	aretis	too **long**
„	aretur	"**longs**" are realizing
„	ari	"longs" are realizing freely
„	arier	"longs" are forcing market
„	at	will **lose** nothing (by)
„	atae	totally lost
„	atarum	are heavy losers
„	atior	without loss
„	atius	no loss
„	atos	owing to heavy losses
„	atote	loss recovered
„	atum	in 1 **lot**
„	aturae	in 2 lots
„	aturi	in several lots
„	aturos	in one or several lots
„	averam	in a **lump** sum
„	averit	**machinery** broken down
„	avero	machinery deranged
„	avimus	by next **mail**
„	avisse	by to-day's mail
„	avisti	received no mail to-day
„	avit	mail steamer
„	emus	to be **made**
„	ent	to be made up
„	es	old make
„	etis	new make
„	etur	from **makers**
„	or	by makers

crem	abam	makers asking
„	abamus	makers are well engaged
„	abant	makers are short of work
„	abare	send **manifest**
„	abaris	as per manifest
„	abatur	wire how **many**
„	abimur	with a **margin** of
„	abis	trade **mark**
„	abitis	all marks
„	abo	other marks
„	amini	usual mark
„	amur	your mark
„	andam	our (my) mark
„	andi	mark as follows
„	andior	to be marked as usual
„	andius	wire what mark
„	andos	wire **market** report
„	ans	wire opinion on the future of market
„	antem	keep us advised of any changes in the market
„	antis	in the open market
„	antium	no change in the market
„	anto	market tends against buyers
„	antur	market tends against sellers
„	aremus	rather more doing
„	arent	there is less doing
„	arere	market active
„	ares	„ more active
„	aretis	„ very active
„	aretur	„ active and strong
„	ari	„ rallying
„	arier	„ „ all round
„	at	„ buoyant tone
„	atae	„ firm undertone
„	atarum	„ strong
„	atior	„ „ and firm
„	atius	„ „ „ dearer
„	atos	„ firm with rather more doing
„	atote	„ firmer and dearer
„	atum	„ tight
„	aturae	„ stagnant
„	aturi	„ flat
„	aturos	„ very flat
„	averam	„ irregular
„	averit	„ lifeless
„	avero	„ neglected
„	avimus	„ quiet
„	avisse	„ quiet but steady
„	avisti	„ very quiet
„	avit	„ more regular
„	emus	„ easy
„	ent	„ easier
„	es	„ slightly easier
„	etis	„ depressed
„	etur	
„	or	

culp	**abam**	market upward tendency
„	**abamus**	market downward tendency
„	**abant**	**mate's** receipt
„	**abare**	against mate's receipt
„	**abaris**	at maturity
„	**abatur**	wire what **maximum**
„	**abimur**	guaranteed maximum
„	**abis**	what do you **mean** by?
„	**abitis**	by all means
„	**abo**	**measurement**
„	**amini**	measurement goods
„	**amur**	at **meeting** of creditors
„	**andam**	as **mentioned**
„	**andi**	in the way mentioned
„	**andior**	lawful **merchandise**
„	**andius**	wire what **minimum**
„	**andos**	guaranteed minimum
„	**ans**	**misc**ellaneous
„	**antem**	owing to a **mistake**
„	**antis**	there must be a mistake
„	**antium**	mistaken, should read:
„	**anto**	**misund**erstanding
„	**antur**	owing to a misunderstanding
„	**aremus**	**mixed** with
„	**arent**	mixed options
„	**arere**	not a **moment** should be lost
„	**ares**	moment seems favourable (for)
„	**aretis**	there is a stringency of **money**
„	**aretur**	there is a plethora of money
„	**ari**	**more** or less
„	**arier**	5% more or less
„	**at**	10 „ „ „ „
„	**atae**	15 „ „ „ „
„	**atarum**	much more
„	**atior**	more than
„	**atius**	not more than
„	**atos**	as **much** as possible
„	**atote**	not much
„	**atum**	**name** unknown
„	**aturae**	wire name
„	**aturi**	wire name and address
„	**aturos**	in our name
„	**averam**	in my „
„	**averit**	in his „
„	**avero**	in your „
„	**avimus**	wire **nature** of
„	**avisse**	**navigation** closed
„	**avisti**	navigation will reopen
„	**avit**	navigation open
„	**emus**	navigation will close
„	**ent**	if **necessary**
„	**es**	not necessary
„	**etis**	wire whether necessary
„	**etur**	it is necessary to
„	**or**	in case of **need**

cune	**abam**	in case of need with
„	**abamus**	you need not
„	**abant**	urgently needed
„	**abare**	do the needful
„	**abaris**	shall do the needful
„	**abatur**	**news** is correct
„	**abimur**	news is contradicted
„	**abis**	no news of
„	**abitis**	all **night**
„	**abo**	during night
„	**amini**	decree **nisi**
„	**amur**	**nominal** being unavailable
„	**andam**	there is **no**
„	**andi**	provided there is no
„	**andior**	there is none
„	**andius**	**North-East**
„	**andos**	North-West
„	**ans**	or **not**
„	**antem**	not yet
„	**antis**	if not
„	**antium**	give **notice**
„	**anto**	given notice
„	**antur**	subject to .... notice
„	**aremus**	**notwith**standing
„	**arent**	not just **now**
„	**arere**	**null** and void
„	**ares**	to be **numbered**
„	**aretis**	wire number
„	**aretur**	not **obtainable**
„	**ari**	we (I) make you a firm **offer** until cancelled
„	**arier**	we (I) make you firm offer for immediate telegr. reply
„	**at**	we (I) make you chance offer for immediate telegr. reply
„	**atae**	try to make firm offer
„	**atarum**	we (I) cancel offer
„	**atior**	we (I) accept firm offer, business will be considered concluded unless you wire at once to the contrary
„	**atius**	this is quite a chance offer
„	**atos**	your offer does not encourage us (me) to make you a counter-offer
„	**atote**	if cannot accept offer try to make counter-offer
„	**atum**	we (I) offer you without prejudice
„	**aturae**	offer is far from being acceptable. you may consider it cancelled
„	**aturi**	with firm offer in hand can probably buy cheaper
„	**aturos**	**omnium** scrip
„	**averam**	and then **on** to
„	**averit**	on and after
„	**avero**	**once** again
„	**avimus**	at once
„	**avisse**	**one** or the other
„	**avisti**	one and the same
„	**avit**	one or two
„	**emus**	at the **opening**
„	**ent**	wire **opinion** on
„	**es**	in our (my) opinion
„	**etis**	according to general opinion
„	**etur**	the general opinion is in favor of
„	**or**	

limit— p. 12

curv	**abam**	cannot express an opinion (on)
„	**abamus**	it is too early to express an opinion (on)
„	**abant**	the opinion prevails that
„	**abare**	take advantage of **opportunity**
„	**abaris**	at the first opportunity
„	**abatur**	try to get **option**
„	**abimur**	buyer's option
„	**abis**	seller's option
„	**abitis**	your option
„	**abo**	owner's option
„	**amini**	our (my) option
„	**amur**	usual option
„	**andam**	declare option
„	**andi**	option to be declared
„	**andior**	in **order**
„	**andius**	or order
„	**andos**	to order
„	**ans**	to our (my) order
„	**antem**	your order
„	**antis**	from receipt of order
„	**antium**	from date of order
„	**anto**	to call for orders at
„	**antur**	until further orders
„	**aremus**	cancelled order
„	**arent**	receiving order
„	**arere**	money order
„	**ares**	send money order
„	**aretis**	repeat order
„	**aretur**	repeat orders
„	**ari**	executed all orders
„	**arier**	order in force
„	**at**	renew order
„	**atae**	trial order
„	**atarum**	order reached too late
„	**atior**	order has been booked
„	**atius**	suspend execution of order
„	**atos**	„ „ „ orders
„	**atote**	suspend execution of order financial position being critical
„	**atum**	suspend execution of order financial position being unfavorable
„	**aturae**	suspend execution of order owing to insolvency of client
„	**aturi**	suspend execution of order failure of client being imminent
„	**aturos**	stopped order
„	**averam**	have no orders on hand
„	**averit**	did you receive order?
„	**avero**	did not receive order
„	**avimus**	received order
„	**avisse**	certificate of **origin**
„	**avisti**	certificates of origin
„	**avit**	send certificate of origin
„	**emus**	wire origin of
„	**ent**	and **other**
„	**es**	or other
„	**etis**	or any other
„	**etur**	one or the other
„	**or**	no other

damn	**abam**	there is another
„	**abamus**	there is no other
„	**abant**	**out** of
„	**abare**	out and out
„	**abaris**	total **output**
„	**abatur**	wire estimate of **outturn**
„	**abimur**	is now practically **over**
„	**abis**	**overdrawn** capital
„	**abitis**	avoid overdrawing
„	**abo**	greatly **overdue**
„	**amini**	overestimated
„	**amur**	**owing** to
„	**andam**	owing probably to
„	**andi**	our (my) **own**
„	**andior**	for a/c of owners
„	**andius**	usual **packing**
„	**andos**	inclusive of packing
„	**ans**	exclusive of packing
„	**antem**	owing to defective packing
„	**antis**	owing to a **panic**
„	**antium**	a state of panic prevails
„	**anto**	**paper** money
„	**antur**	forced paper currency
„	**aremus**	at **par**
„	**arent**	below par
„	**arere**	above par
„	**ares**	par value
„	**aretis**	on the par amount
„	**aretur**	about on a par with
„	**ari**	all business **paralized**
„	**arier**	by **parcel** post
„	**at**	post 1 parcel
„	**atae**	„ 2 parcels
„	**atarum**	„ 3 „
„	**atior**	„ 4 „
„	**atius**	„ 5 „
„	**atos**	only **part** of
„	**atote**	on our (my) part
„	**atum**	on your part
„	**aturae**	on their (his) part
„	**aturi**	all or any part of
„	**aturos**	to take part (in)
„	**averam**	taking part (in)
„	**averit**	to all parts
„	**avero**	from all parts
„	**avimus**	wire **particulars** (re)
„	**avisse**	particulars by letter
„	**avisti**	no further particulars received
„	**avit**	admitted as a **partner**
„	**emus**	partnership dissolved
„	**ent**	a **party to**
„	**es**	all parties concerned
„	**etis**	inform all parties concerned
„	**etur**	on **passage**
„	**or**	outward passage

dat	**abam**	1st class passage
„	**abamus**	2nd „ „
„	**abant**	3rd „ „
„	**abare**	book 1st class passage
„	**abaris**	book 2nd „ „
„	**abatur**	passenger steamer
„	**abimur**	1st class passenger(s)
„	**abis**	2nd „ „
„	**abitis**	passengers and goods
„	**abo**	passengers and crew
„	**amini**	**patent** No.
„	**amur**	**pay** for our a/c
„	**andam**	„ „ my a/c
„	**andi**	pay bonus
„	**andior**	pay interest
„	**andius**	paid interest
„	**andos**	amount paid
„	**ans**	wire amount paid
„	**antem**	wire amount to be paid
„	**antis**	to be paid
„	**antium**	to be paid at once
„	**anto**	part paid
„	**antur**	fully paid
„	**aremus**	all paid up
„	**arent**	pay day
„	**arere**	make payable
„	**ares**	wire as soon as paid
„	**aretis**	in part payment of
„	**aretur**	for want of payment
„	**ari**	enforce payment
„	**arier**	against payment (of)
„	**at**	suspended payments
„	**atae**	likely to suspend payments
„	**atarum**	resumed payments
„	**atior**	present for payment
„	**atius**	payment cash
„	**atos**	„ „ 2% discount
„	**atote**	„ „ $2^{1}/_{2}$% „
„	**atum**	„ „ 3% „
„	**aturae**	„ „ with order
„	**aturi**	„ „ on delivery
„	**aturos**	„ $^{2}/_{3}$ cash with order
„	**averam**	payment $^{2}/_{3}$ cash with order and balance
„	**averit**	payment $^{2}/_{3}$ cash with order and balance on delivery
„	**avero**	payment $^{3}/_{4}$ cash with order
„	**avimus**	payment $^{3}/_{4}$ cash with order and balance
„	**avisse**	payment $^{3}/_{4}$ cash with order and balance on delivery
„	**avisti**	repayable at par
„	**avit**	repayable by instalments
„	**emus**	apply for **permission**
„	**ent**	to take **place**
„	**es**	wire when to take place
„	**etis**	will not take place
„	**etur**	**plague** broken out (here)
„	**or**	free from plague

deam	**abam**	**Plaintiff**(s) to pay costs
„	**abamus**	in favor of Plaintiff(s)
„	**abant**	against Plaintiff(s)
„	**abare**	**point** out that
„	**abaris**	owing to **political** rumours
„	**abatur**	in **port**
„	**abimur**	loading port
„	**abis**	port of discharge
„	**abitis**	safe port
„	**abo**	all **positions**
„	**amini**	other positions
„	**amur**	forward positions
„	**andam**	unsold position
„	**andi**	oversold position
„	**andior**	wire unsold position
„	**andius**	wire oversold position
„	**andos**	wire opinion on position (of)
„	**ans**	position is excellent
„	**antem**	„ is satisfactory
„	**antis**	„ has improved
„	**antium**	„ „ „ considerably
„	**anto**	„ is less satisfactory
„	**antur**	„ „ unfavorable
„	**aremus**	„ „ uncertain
„	**arent**	„ „ critical
„	**arere**	„ „ growing critical
„	**ares**	„ has become critical
„	**aretis**	„ is unchanged
„	**aretur**	in a critical position
„	**ari**	owing to unfavorable position
„	**arier**	owing to critical position
„	**at**	if **possible**
„	**atae**	if not possible
„	**atarum**	if not possible, wire
„	**atior**	wire whether possible
„	**atius**	as far as possible
„	**atos**	as much as possible
„	**atote**	as little as possible
„	**atum**	as large as possible
„	**aturae**	it is possible (that)
„	**aturi**	possible but not probable
„	**aturos**	whenever possible
„	**averam**	or if possible
„	**averit**	by **post**
„	**avero**	**postponed** till
„	**avimus**	send full **power**
„	**avisse**	sent full power
„	**avisti**	send necessary power to act in [your name
„	**avit**	take **precautions**
„	**emus**	taken precautions
„	**ent**	all the **preceding**
„	**es**	this applies to all preceding
„	**etis**	**preferential** claim
„	**etur**	
„	**or**	

dens	**abam**	preferential claims
„	**abamus**	preference stock
„	**abant**	wire whether prefer
„	**abare**	wire which you prefer
„	**abaris**	deem it preferable (to)
„	**abatur**	if you prefer
„	**abimur**	give preference to
„	**abis**	given preference to
„	**abitis**	in preference to
„	**abo**	without **prejudice**
„	**amini**	greatly prejudiced (by)
„	**amur**	**preliminary**
„	**andam**	at a **premium**
„	**andi**	wire premium
„	**andior**	are **prep**ared to
„	**andius**	prepayment
„	**andos**	against prepayment
„	**ans**	against prepayment of the amount
„	**antem**	at **present**
„	**antis**	not for the present
„	**antium**	in presence of
„	**anto**	present and future
„	**antur**	up to the present
„	**aremus**	on presentation
„	**arent**	to be presented
„	**arere**	owing to **pressure**
„	**ares**	by false **pretences**
„	**aretis**	**prevent** by all possible means
„	**aretur**	wire best possible **price** for
„	**ari**	wire best possible prices for
„	**arier**	at acceptable prices
„	**at**	„ actual price
„	**atae**	„ „ prices
„	**atarum**	„ any price
„	**atior**	„ approximate price (of)
„	**atius**	„ price asked
„	**atos**	„ prices asked
„	**atote**	„ average price
„	**atum**	„ „ prices
„	**aturae**	„ chance price
„	**aturi**	„ „ prices
„	**aturos**	„ competitors' prices
„	**averam**	„ contract price
„	**averit**	„ „ prices
„	**avero**	„ cost price
„	**avimus**	„ „ prices
„	**avisse**	„ exorbitant price
„	**avisti**	„ „ prices
„	**avit**	„ favorable prices
„	**emus**	„ very favorable prices
„	**ent**	„ fixed prices
„	**es**	„ high price
„	**etis**	„ higher price
„	**etur**	„ high prices
„	**or**	„ higher prices

dict	**abam**	at very high prices
„	**abamus**	„ very low prices
„	**abant**	„ prices offered
„	**abare**	„ previous price
„	**abaris**	„ „ prices
„	**abatur**	„ prohibitive prices
„	**abimur**	„ reduced price
„	**abis**	„ „ prices
„	**abitis**	„ same price (as)
„	**abo**	„ „ prices (as)
„	**amini**	„ settlement price
„	**amur**	„ „ prices
„	**andam**	„ these prices
„	**andi**	„ this price
„	**andior**	„ unremunerative prices
„	**andius**	„ usual price
„	**andos**	„ „ prices
„	**ans**	this is a chance price
„	**antem**	these are chance prices
„	**antis**	contract price
„	**antium**	„ prices
„	**anto**	cost price
„	**antur**	„ prices
„	**aremus**	below cost price
„	**arent**	„ „ prices
„	**arere**	purchase price
„	**ares**	sale price
„	**aretis**	settlement price
„	**aretur**	„ prices
„	**ari**	price is barely covering cost
„	**arier**	prices are barely covering cost
„	**at**	price is exorbitant
„	**atae**	prices are exorbitant
„	**atarum**	price is very favorable
„	**atior**	prices are very favorable
„	**atius**	prices are unremunerative
„	**atos**	wire market price
„	**atote**	„ „ prices
„	**atum**	„ approximate price
„	**aturae**	„ „ prices
„	**aturi**	„ competitors' price
„	**aturos**	„ „ prices
„	**averam**	„ price for large quantity
„	**averit**	„ prices for large quantities
„	**avero**	„ price offered
„	**avimus**	„ prices offered
„	**avisse**	wire best possible price and shortest possible delay of delivery for
„	**avisti**	wire price for A I quality
„	**avit**	„ „ „ choice quality
„	**emus**	„ „ „ superior „
„	**ent**	„ „ „ quality guaranteed
„	**es**	„ „ „ current quality
„	**etis**	„ „ „ middling „
„	**etur**	„ „ „ ordinary „
„	**or**	

dign abam		wire price for good quality
„ **abamus**		wire price for other quality
„ **abant**		wire prices for other qualities
„ **abare**		wire price for usual quality
„ **abaris**		wire price for quality = to sample
„ **abatur**		wire price for same quantity and quality as last supplied
„ **abimur**		wire settlement price
„ **abis**		wire settlement prices
„ **abitis**	prices are higher owing to	scanty arrivals
„ **abo**		bears taking profit
„ **amini**		bulls buying
„ **amur**		large buying
„ **andam**		buying orders
„ **andi**		repurchases
„ **andior**		demand to cover
„ **andius**		unfavorable crop reports
„ **andos**		crop damage
„ **ans**		strong demand
„ **antem**		increasing demand
„ **antis**		higher freights
„ **antium**		small production
„ **anto**		dearer raw material
„ **antur**		small receipts
„ **aremus**		scarcity of sellers
„ **arent**		small shipments
„ **arere**		small stocks
„ **ares**		small supplies
„ **aretis**	prices are lower owing to	unfavorable advices
„ **aretur**		bearish advices
„ **ari**		large arrivals
„ **arier**		better crop outlook
„ **at**		slack demand
„ **atae**		absence of demand
„ **atarum**		lower freights
„ **atior**		liquidations
„ **atius**		large offerings
„ **atos**		large production
„ **atote**		realization sales
„ **atum**		heavy selling
„ **aturae**		pressure to sell
„ **aturi**		resales
„ **aturos**		heavy shipments
„ **averam**		large stocks
„ **averit**		increasing stocks
„ **avero**		large supplies
„ **avimus**		increasing supplies
„ **avisse**		prices are understood
„ **avisti**		wire how prices are understood
„ **avit**		list prices
„ **emus**		at prices of old list
„ **ent**		at prices of new list
„ **es**		prices nominal, no buyers
„ **etis**		owing to reduction of prices
„ **etur**		repeat price in plain figures
„ **or**		repeat prices in plain figures

don abam	prices are in favour of buyers
„ **abamus**	„ „ „ „ „ sellers
„ **abant**	prices are above buyers' views
„ **abare**	in **principle**
„ **abaris**	this is strictly **private**
„ **abatur**	in all **probability**
„ **abimur**	it is probable (that)
„ **abis**	owing to probable
„ **abitis**	hardly probable
„ **abo**	**proceed** at once
„ **amini**	to proceed at once
„ **amur**	will be proceeded against
„ **andam**	stay legal proceedings
„ **andi**	net proceeds
„ **andior**	out of net proceeds
„ **andius**	wire net proceeds
„ **andos**	costs of procedure
„ **ans**	estimate costs of procedure
„ **antem**	per **procuration**
„ **antis**	will sign per procuration
„ **antium**	total **production**
„ **anto**	total production of season
„ **antur**	wire estimate of production
„ **aremus**	overproduction
„ **arent**	owing to overproduction
„ **arere**	without **profit**
„ **ares**	with a profit (of)
„ **aretis**	profit realized
„ **aretur**	not leaving any profit
„ **ari**	profit and loss a/c
„ **arier**	**progressing** favorably
„ **at**	**promissory** note
„ **atae**	carry on energetic **prop**aganda
„ **atarum**	movable property
„ **atior**	immovable property
„ **atius**	in **proportion** to
„ **atos**	in equal proportion
„ **atote**	to be proportionate to
„ **atum**	if cannot accept **proposal** make counter-proposal
„ **aturae**	as proposed
„ **aturi**	accept proposal
„ **aturos**	there is every **prospect** (of)
„ **averam**	there is no prospect (of)
„ **averit**	prospects are excellent
„ **avero**	prospects are encouraging
„ **avimus**	prospects are unfavorable
„ **avisse**	prospects are hopeless
„ **avisti**	send prospectus
„ **avit**	**protest** against
„ **emus**	lodge protest
„ **ent**	lodged protest
„ **es**	we (I) protest against
„ **etis**	„ „ protested against
„ **etur**	under protest
„ **or**	owing to protest

dur	**abam**	accept under protest
„	**abamus**	avoid protest
„	**abant**	bill protested for non-acceptance
„	**abare**	bill protested for non-payment
„	**abaris**	provided there is
„	**abatur**	„ you
„	**abimur**	„ they
„	**abis**	wire what **quality**
„	**abitis**	wire opinion on quality
„	**abo**	according to quality
„	**amini**	if quality is satisfactory
„	**amur**	off quality
„	**andam**	quality and condition
„	**andi**	allowance for deficiency of quality
„	**andior**	quality very deficient
„	**andius**	quality is not equal to sample
„	**andos**	goods refused for deficiency of quality
„	**ans**	owing to bad quality
„	**antem**	all qualities
„	**antis**	all qualities and descriptions
„	**antium**	quality = to
„	**anto**	quality about = to
„	**antur**	quality = to sample
„	**aremus**	quality about = to sample
„	**arent**	best quality procurable
„	**arere**	choice quality
„	**ares**	choice qualities
„	**aretis**	superior quality
„	**aretur**	superior qualities
„	**ari**	quality guaranteed
„	**arier**	quality guaranteed = to
„	**at**	good quality
„	**atae**	good qualities
„	**atarum**	current quality
„	**atior**	ordinary quality
„	**atius**	middling quality
„	**atos**	cheaper quality
„	**atote**	inferior quality
„	**atum**	usual quality
„	**aturae**	same quality as
„	**aturi**	same quality as last supplied
„	**aturos**	send same quality as last supplied
„	**averam**	send same quality and on same terms as last supplied
„	**averit**	send same quality, same quantity and on same terms as last supplied
„	**avero**	wire **quantity**
„	**avimus**	approximate quantity
„	**avisse**	maximum quantity
„	**avisti**	minimum „
„	**avit**	total „
„	**emus**	large „
„	**ent**	larger „
„	**es**	small „
„	**etis**	smaller „
„	**etur**	option double quantity
„	**or**	increase quantity (to)

ebri	**abam**	try to increase quantity (to)
„	**abamus**	smaller quantities comparatively dearer
„	**abant**	larger quantities comparatively cheaper
„	**abare**	wire approximate quantity
„	**abaris**	„ maximum „
„	**abatur**	„ minimum „
„	**abimur**	„ total quantity
„	**abis**	out of quantity
„	**abitis**	out of total quantity
„	**abo**	detained by **quarantine**
„	**amini**	owing to quarantine
„	**amur**	subject to quarantine
„	**andam**	quarantine removed
„	**andi**	on **quay**
„	**andior**	ex quay
„	**andius**	sold from quay in transit
„	**andos**	in **question**
„	**ans**	question in dispute
„	**antem**	question is settled
„	**antis**	wire **quotation** (for)
„	**antium**	wire quotations (for)
„	**anto**	quotations nominal no buyers
„	**antur**	**rain** is urgently wanted
„	**aremus**	owing to rain
„	**arent**	no rain has fallen
„	**arere**	at the **rate** of
„	**ares**	wire rate of
„	**aretis**	at a reduced rate
„	**aretur**	at reduced rates
„	**ari**	expect a **reaction**
„	**arier**	to be ready
„	**at**	will be **ready**
„	**atae**	if not ready
„	**atarum**	owing to **realizations**
„	**atior**	realize to take profits
„	**atius**	amount realized
„	**atos**	wire amount realized
„	**atote**	out of amount realized
„	**atum**	sold and realized
„	**aturae**	there is no **reason** for
„	**aturi**	wire what **rebate**
„	**aturos**	no rebate
„	**averam**	5 % rebate
„	**averit**	10 % „
„	**avero**	15 % „
„	**avimus**	20 % „
„	**avisse**	25 % „
„	**avisti**	30 % „
„	**avit**	$33\frac{1}{3}$ % „
„	**emus**	40 % „
„	**ent**	45 % „
„	**es**	50 % „
„	**etis**	against receipt
„	**etur**	receipts in the interior
„	**or**	on receipt of

edol	**abam**	amount **recovered**
„	**abamus**	to be **reduced**
„	**abant**	against reduction (of)
„	**abare**	state **references**
„	**abaris**	owing to **refusal**
„	**abatur**	in case of refusal
„	**abimur**	may be **regarded** as
„	**abis**	**reimbursement**
„	**abitis**	reimbursements
„	**abo**	for reimbursement
„	**amini**	avoid **relations** with
„	**amur**	advise relatives
„	**andam**	at what rate can you **relet**?
„	**andi**	thouroughly **reliable**
„	**andior**	**remit** by telegraphic transfer
„	**andius**	remitted by „ „
„	**andos**	remit all available funds
„	**ans**	drawing is more advantageous at present than remitting
„	**antem**	remitting is more advantageous at present than drawing
„	**antis**	to be remitted
„	**antium**	**renewed** till
„	**anto**	exhausted, renew
„	**antur**	owing to **repairs**
„	**aremus**	**repeat** of your message the word following the word
„	**arent**	repeat of your message all the words following the word
„	**arere**	repeat in plain words
„	**ares**	to be replaced by
„	**aretis**	**reply** at once by wire
„	**aretur**	cannot get a reply
„	**ari**	according to **reports** (from)
„	**arier**	it is reported that
„	**at**	reports require confirmation
„	**atae**	reports are conflicting
„	**atarum**	report has proved correct
„	**atior**	send full report
„	**atius**	**represented** (by)
„	**atos**	representing
„	**atote**	on representation
„	**atum**	at the **request** of
„	**aturae**	**requirements**
„	**aturi**	wire your requirements
„	**aturos**	do you require further?
„	**averam**	require further
„	**averit**	for our (my) requirements
„	**avero**	against your requirements
„	**avimus**	„ our (my) requirements
„	**avisse**	„ urgent requirements
„	**avisti**	**reserve** fund
„	**avit**	under all reserves
„	**emus**	in every **respect**
„	**ent**	**responsibility**
„	**es**	on our own responsibility
„	**etis**	on my own responsibility
„	**etur**	on your own responsibility
„	**or**	**restrict** as much as possible

educ	**abam**	wire **result**
„	**abamus**	wire probable result
„	**abant**	result is satisfactory
„	**abare**	result is highly satisfactory
„	**abaris**	result is unsatisfactory
„	**abatur**	result is most unsatisfactory
„	**abimur**	expect favorable result
„	**abis**	result still unknown
„	**abitis**	**retail** business
„	**abo**	**return** document(s) sent
„	**amini**	all **right**
„	**amur**	owing to a **ring**
„	**andam**	a ring has been formed
„	**andi**	risk of failure
„	**andior**	at the risk of
„	**andius**	at your risk
„	**andos**	at our (my) risk
„	**ans**	at their (his) risk
„	**antem**	do you expect there will be any [risk]
„	**antis**	we (I) do not expect there will be any [risk]
„	**antium**	it would be too risky (to)
„	**anto**	no **room** available
„	**antur**	by the quickest **route**
„	**aremus**	by the cheapest route
„	**arent**	there is a **rumour** that
„	**arere**	owing to rumours regarding
„	**ares**	owing to an unconfirmed rumour (of)
„	**aretis**	rumour requires confirmation
„	**aretur**	there is a persistent rumour (of)
„	**ari**	no credence is given to the rumour
„	**arier**	rumour entirely unfounded
„	**at**	according to a rumour
„	**atae**	it is **said** that
„	**atarum**	said to be
„	**atior**	owing to large **sales**
„	**atius**	„ „ realization sales
„	**atos**	wire at what price you can sell
„	**atote**	„ „ „ „ „ recommend selling
„	**atum**	wire quantity and price you can sell
„	**aturae**	wire quantity and price you recommend selling
„	**aturi**	sell for our (my) a/c
„	**aturos**	sold for your a/c
„	**averam**	to be sold
„	**averit**	provided unsold
„	**avero**	still unsold
„	**avimus**	only saleable at
„	**avisse**	this is a chance sale
„	**avisti**	against sales
„	**avit**	against sales in anticipation
„	**emus**	sell as much as possible
„	**ent**	selling in anticipation
„	**es**	all sold
„	**etis**	out and out sale
„	**etur**	part sold
„	**or**	sold nothing

prot.– p. 18

ejer	abam	cover short sales
„	abamus	sales are satisfactory continue
„	abant	sell in the most advantageous market
„	abare	buying is more advantageous than selling at present
„	abaris	selling is more advantageous than buying at present
„	abatur	the **same as**
„	abimur	send **sample**
„	abis	send samples
„	abitis	sent sample
„	abo	sent samples
„	amini	send larger sample
„	amur	send turn-out sample
„	andam	„ counter-sample
„	andi	as per sample
„	andior	as per your sample(s)
„	andius	„ „ our (my) „
„	andos	sample(s) did not reach
„	ans	send sample taken at random
„	antem	sent „ „ „ „
„	antis	send samples of all qualities
„	antium	„ „ „ „ articles [for
„	anto	„ sample and quote lowest price
„	antur	„ samples „ „ „ prices
„	aremus	quality not = to sample [for
„	arent	quality not = to sample, goods
„	arere	received sample(s) [refused
„	ares	**satisfactory**
„	aretis	highly satisfactory
„	aretur	very **scarce**
„	ari	rather scarce
„	arier	owing to scarcity (of)
„	at	per **score**
„	atae	**scrip** certificate
„	atarum	„ certificates
„	atior	„ coupon
„	atius	against **security**
„	atos	keep this **secret**
„	atote	it **seems**
„	atum	it does not seem
„	aturae	**settlement** through
„	aturi	settling day
„	aturos	in settlement of
„	averam	„ „ „ balance
„	averit	by amicable settlement
„	avero	**sharing** expenses
„	avimus	**ship** as much as possible
„	avisse	„ „ soon „ „
„	avisti	shall ship as soon as possible
„	avit	to be shipped
„	emus	against shipment
„	ent	„ shipments
„	es	ship and cargo
„	etis	in what shortest delay can you ship?
„	etur	late shipped
„	or	being late shipped to pay allowance (of)

elev	abam	short shipped
„	abamus	on "**short**" a/c
„	abant	**should** not
„	abare	it should be
„	abaris	long **sight**
„	abatur	short sight
„	abimur	longer sight
„	abis	shorter sight
„	abitis	8 days' sight
„	abo	30 „ „
„	amini	60 „ „
„	amur	3 months' sight
„	andam	4 „ „
„	andi	6 „ „
„	andior	to be **signed**
„	andius	there are signs (of)
„	andos	there are no signs (of)
„	ans	signature is not authentic
„	antem	**similar** to
„	antis	**simultaneous**(ly)
„	antium	**since** last
„	anto	wire re **situation**
„	antur	situation is favorable
„	aremus	„ „ more favorable
„	arent	„ „ less favorable
„	arere	„ „ unfavorable
„	ares	„ „ critical
„	aretis	„ „ unchanged
„	aretur	„ „ improving
„	ari	wire **size**
„	arier	very **small**
„	at	too small
„	atae	owing to **snowstorms**
„	atarum	very **soon**
„	atior	as soon as possible
„	atius	as soon as you consider it advisable
„	atos	sooner if possible
„	atote	sooner or later
„	atum	all **sorts** (of)
„	aturae	all sorts and qualities
„	aturi	from reliable **source**
„	aturos	hear from trustworthy source
„	averam	**South**-East
„	averit	South-West
„	avero	can you **spare**?
„	avimus	how much can you spare?
„	avisse	cannot spare anything
„	avisti	in **specie**
„	avit	**specification**
„	emus	as per specification
„	ent	send specification
„	es	inclined to **speculate**
„	etis	speculation
„	etur	on speculation
„	or	

elix	**abam**	for speculation
„	**abamus**	owing to speculations
„	**abant**	„ „ reckless speculation
„	**abare**	speculative
„	**abaris**	speculator/s
„	**abatur**	on the **spot**
„	**abimur**	of old **standing**
„	**abis**	wire **state** of
„	**abitis**	send statement
„	**abo**	there is no **steamer** at present
„	**amini**	there is no other steamer at
„	**amur**	wire name of steamer [present
„	**andam**	take energetic **steps**
„	**andi**	taken necessary steps
„	**andior**	shall take necessary steps
„	**andius**	as **stipulated**
„	**andos**	to be expressly stipulated
„	**ans**	no **stock**
„	**antem**	out of stock
„	**antis**	surplus stocks
„	**antium**	stock all cleared
„	**anto**	accumulate stocks
„	**antur**	accumulating stocks
„	**aremus**	stock nearly exhausted
„	**arent**	stocktaking
„	**arere**	owing to bad **stowage**
„	**ares**	**stranded** on rocks
„	**aretis**	owing to **strike**
„	**aretur**	strike is imminent
„	**ari**	strike is ended
„	**arier**	under the **style** of
„	**at**	under the same style
„	**atae**	on the **subject**
„	**atarum**	subject to
„	**atior**	to be **submitted** (to)
„	**atius**	**subscription**
„	**atos**	what can you **substitute**?
„	**atote**	as a substitute for
„	**atum**	to be substituted
„	**aturae**	to be **succeeded** (by)
„	**aturi**	proved a great **success**
„	**aturos**	successful/ly
„	**averam**	very successful/ly
„	**averit**	**such** as
„	**avero**	there is no such
„	**avimus**	**suffer**ing severely from
„	**avisse**	wire whether **sufficient**
„	**avisti**	if sufficient
„	**avit**	if not sufficient
„	**emus**	not sufficient/ly
„	**ent**	take out a **summons** (against)
„	**es**	taken „ „ „ „
„	**etis**	to be **supplied**
„	**etur**	new supplies
„	**or**	out of supplies

eloc	**abam**	visible supply
„	**abamus**	available **surplus**
„	**abant**	out of surplus
„	**abare**	apply surplus (to)
„	**abaris**	entirely **suspended**
„	**abatur**	temporarily suspended
„	**abimur**	remaining in suspense (till)
„	**abis**	on old **system**
„	**abitis**	on new system
„	**abo**	will **take** about
„	**amini**	new **tariff**
„	**amur**	old tariff
„	**andam**	on basis of old tariff
„	**andi**	„ „ „ new tariff
„	**andior**	new tariff will come into oper-
„	**andius**	**telegraph** whether [ation
„	**andos**	„ as soon as possible
„	**ans**	„ at once
„	**antem**	your telegram arrived too late
„	**antis**	telegraphically
„	**antium**	telegraphic transfer
„	**anto**	„ transfers
„	**antur**	by telegraphic transfer
„	**aremus**	according to a telegram from
„	**arent**	shall telegraph later on
„	**arere**	„ „ further later on
„	**ares**	telegraph reply
„	**aretis**	„ this to
„	**aretur**	only **temporary**
„	**ari**	**tenders** are invited for
„	**arier**	apply for tenders
„	**at**	your tenders are accepted
„	**atae**	lowest tenders
„	**atarum**	accept **terms** mentioned
„	**atior**	on best possible terms
„	**atius**	wire best possible terms for
„	**atos**	special terms (for)
„	**atote**	on more favorable terms
„	**atum**	on less favorable terms
„	**aturae**	terms are inacceptable
„	**aturi**	on usual terms
„	**aturos**	„ new terms
„	**averam**	„ old terms
„	**averit**	terms should be clearly stipul-
„	**avero**	**thence** to [ated
„	**avimus**	**there** is
„	**avisse**	„ will be
„	**avisti**	„ was
„	**avit**	**through**rate
„	**emus**	wire throughrate for
„	**ent**	on through B/L
„	**es**	in a very short **time**
„	**etis**	for some time to come
„	**etur**	for the time being
„	**or**	within which time?

emut	**abam**	there is still **time**
„	**abamus**	„ „ no time to be lost
„	**abant**	„ „ „ more time (for)
„	**abare**	owing to lack of time
„	**abaris**	is there still time (for)?
„	**abatur**	only a question of time
„	**abimur**	**tonnage** unavailable
„	**abis**	wire **total**
„	**abitis**	out of total
„	**abo**	to be **transferred** (to)
„	**amini**	in **transit**
„	**amur**	**transshipment** (at)
„	**andam**	transshipped (at)
„	**andi**	transshipping
„	**andior**	without transshipment
„	**andius**	we are (I am) in **treaty** for
„	**andos**	on **trial**
„	**ans**	recommend a trial
„	**antem**	is it **true** that?
„	**antis**	as per **type**
„	**antium**	send type
„	**anto**	send types
„	**antur**	owing to a **typhoon**
„	**aremus**	name an **umpire**
„	**arent**	to act as umpire
„	**arere**	**unavailable**
„	**ares**	unavailable for want of sellers
„	**aretis**	if unavailable
„	**aretur**	still unavailable
„	**ari**	owing to **uncertainty** (re)
„	**arier**	not **under**
„	**at**	if under
„	**atae**	or under
„	**atarum**	underestimated
„	**atior**	cannot understand, explain
„	**atius**	{have understood as follows, if correct no reply needed
„	**atos**	will undertake not to
„	**atote**	to undertake not to
„	**atum**	underwriters
„	**aturae**	very **unfavorable**
„	**aturi**	very **unlikely**
„	**aturos**	**unnecessary**
„	**averam**	if unnecessary
„	**averit**	**unremunerative**
„	**avero**	utterly **unsuitable** (for)
„	**avimus**	**until** when?
„	**avisse**	**up to** and including
„	**avisti**	fully up to
„	**avit**	matter is extremely **urgent**
„	**emus**	against **us**
„	**ent**	by us
„	**es**	on us
„	**etis**	from us
„	**etur**	for us
„	**or**	to us

enat	**abam**	as **usual**
„	**abamus**	ad **valorem**
„	**abant**	wire **value**
„	**abare**	approximate value
„	**abaris**	wire approximate value
„	**abatur**	value declared
„	**abimur**	total value
„	**abis**	wire total value
„	**abitis**	or **vice** versa
„	**abo**	**vis** major
„	**amini**	**wait** till
„	**amur**	shall wait till
„	**andam**	for **want** of
„	**andi**	urgently wanted
„	**andior**	in **warehouse**
„	**andius**	ex warehouse
„	**andos**	warehousing expenses
„	**ans**	goods to be warehoused
„	**antem**	warehouse room
„	**antis**	„ room unavailable
„	**antium**	„ rent
„	**anto**	**warn** you against
„	**antur**	in the same **way**
„	**aremus**	„ „ usual way
„	**arent**	„ „ way mentioned
„	**arere**	on the way back
„	**ares**	very **weak**
„	**aretis**	owing to favorable **weather**
„	**aretur**	„ „ unfavorable „
„	**ari**	weather permitting
„	**arier**	first **week** of
„	**at**	1 week
„	**atae**	2 weeks
„	**atarum**	3 „
„	**atior**	4 „
„	**atius**	5 „
„	**atos**	6 „
„	**atote**	wire **weight** (of)
„	**atum**	loss in weight
„	**aturae**	owing to loss in weight
„	**aturi**	there is heavy loss in weight
„	**aturos**	excess weight
„	**averam**	owing to excess weight
„	**averit**	allowance for loss in weight
„	**avero**	demanding allowance for loss [in weight
„	**avimus**	gross weight
„	**avisse**	net weight
„	**avisti**	maximum weight
„	**avit**	minimum weight
„	**emus**	approximate weight
„	**ent**	total weight
„	**es**	wire gross weight
„	**etis**	„ net weight
„	**etur**	„ maximum weight
„	**or**	„ minimum weight

enud	**abam**	wire approximate weight
„	**abamus**	„ total weight
„	**abant**	„ average weight
„	**abare**	„ gross and net weight (of)
„	**abaris**	state gross and net weight of each package
„	**abatur**	doing **well**
„	**abimur**	very well
„	**abis**	fairly well
„	**abitis**	at the **wharves**
„	**abo**	wire **what**
„	**amini**	wire **when**
„	**amur**	uncertain when
„	**andam**	**whereabouts** inascertainable
„	**andi**	wire whereabouts (of)
„	**andior**	of **which**
„	**andius**	all of which
„	**andos**	think it worth **while**
„	**ans**	do not think it worth while
„	**antem**	wire **who**
„	**antis**	who is (are)
„	**antium**	**whole** or part
„	**anto**	wholesale dealer/s
„	**antur**	to **whom**
„	**aremus**	for whom
„	**arent**	wire **why**
„	**arere**	by wire
„	**ares**	**with** or without
„	**aretis**	**working** day and night
„	**aretur**	working capital
„	**ari**	in working order
„	**arier**	work progressing
„	**at**	short of work
„	**atae**	going from bad to **worse**
„	**atarum**	much worse
„	**atior**	not **worth** much
„	**atius**	not worth more than
„	**atos**	absolutely worthless
„	**atote**	it **would**
„	**atum**	it would not
„	**aturae**	would you?
„	**aturi**	would not
„	**aturos**	vessel totally **wrecked** (at)
„	**averam**	1 **year**
„	**averit**	2 years
„	**avero**	3 „
„	**avimus**	4 „
„	**avisse**	half yearly
„	**avisti**	for half year ended
„	**avit**	not **yet**
„	**emus**	if not yet
„	**ent**	estimate **yield**
„	**es**	by **you**
„	**etis**	of „
„	**etur**	on „
„	**or**	to „

err	abam	par suite d'**abondance** de
„	abamus	d'**abord**
„	abant	en l'**absence** de
„	abare	il est préférable de s'**abstenir** complètement
„	abaris	sur place on estime généralement qu'il vaut mieux s'abstenir complètement
„	abatur	**acceptation**
„	abimur	faute d'acceptation
„	abis	contre acceptation
„	abitis	à l'acceptation
„	abo	télégraphiez si vous acceptez
„	amini	un **accident** s'est produit
„	amur	par suite d'accident
„	andam	il n'y a pas d'accident
„	andi	d'**accord**
„	andior	télégraphiez si d'accord
„	andius	si d'accord
„	andos	pas d'accord
„	ans	si pas d'accord
„	antem	„ „ „ télégraphiez
„	antis	pourvu que vous soyez d'accord
„	antium	**achetez** pour notre (mon) compte
„	anto	achetez pour compte à demi
„	antur	télégraphiez à quel prix vous pouvez acheter
„	aremus	télégraphiez à quel prix vous recommandez d'acheter
„	arent	télégraphiez combien et à quel prix vous pouvez acheter
„	arere	télégraphiez combien et à quel prix vous recommandez d'acheter
„	ares	ne vous pressez pas d'acheter
„	aretis	achat d'occasion
„	aretur	en cas d'achat
„	ari	achetez et expédiez
„	arier	il est difficile d'acheter
„	at	il y a **acheteurs**
„	atae	il y a **vendeurs**
„	atarum	il n'y a pas d'acheteurs
„	atior	il n'y a pas de vendeurs
„	atius	il y a beaucoup d'acheteurs
„	atos	il y a beaucoup de vendeurs
„	atote	il y a peu d'acheteurs
„	atum	il y a peu de vendeurs
„	aturae	il n'y a presque pas d'acheteurs
„	aturi	il n'y a presque pas de vendeurs
„	aturos	nous sommes acheteurs
„	averam	nous sommes vendeurs
„	averit	nous ne sommes pas acheteurs
„	avero	nous ne sommes pas vendeurs
„	avimus	à la charge des acheteurs
„	avisse	à la charge des vendeurs
„	avisti	les acheteurs sont réservés
„	avit	vendeurs sont réservés
„	emus	les acheteurs sont empressés
„	ent	les vendeurs sont empressés
„	es	il y a acheteurs mais pas de vendeurs
„	etis	
„	etur	
„	or	

evir	abam	il y a vendeurs mais pas d'acheteurs
„	abamus	**acompte** (sur)
„	abant	à titre d'acompte
„	abare	**actions** nominatives
„	abaris	actions au porteur
„	abatur	en actions
„	abimur	actionnaire/s
„	abis	**adjudication**
„	abitis	adjudicataire/s
„	abo	déclaré/s adjudicataire/s
„	amini	**administrateur/s**
„	amur	administration
„	andam	télégraphiez **adresse**
„	andi	adresse inexacte
„	andior	à votre adresse
„	andius	à notre(mon)adresse
„	andos	adressez correspondance et télégrammes
„	ans	adressez aux soins de
„	antem	adressez poste restante
„	antis	télégraphiez **agio** sur l'or
„	antium	agio sur l'or
„	anto	il s'**agit** de
„	antur	pour être **agréable/s** au/x client/s
„	aremus	ou **ailleurs**
„	arent	il en est **ainsi**
„	arere	s'il en est ainsi
„	ares	télégraphiez s'il en est ainsi
„	aretis	**ajourné** à un temps indéterminé
„	aretur	**amortissable**
„	ari	par **an**
„	arier	1 an
„	at	2 ans
„	atae	3 „
„	atarum	4 „
„	atior	avant nouvel-an
„	atius	après nouvel-an
„	atos	télégraphiez résultat **analyse**
„	atote	selon toute **apparence**
„	atum	**appartenant** à
„	aturae	Court d'**Appel**
„	aturi	**applicable/s** à
„	aturos	à l'**approche** de
„	averam	approvisionnements
„	averit	approvisionné/s
„	avero	bien approvisionné/s
„	avimus	mal „
„	avisse	suffisamment approvisionné/s
„	avisti	nous sommes (je suis) suffisamment approvisionné/s
„	avit	pièces à l'**appui**
„	emus	**approximatif**
„	ent	approximativement
„	es	**argent** en barres
„	etis	**arrangement**
„	etur	arrangement provisoire
„	or	en vertu de l'arrangement

exor	**abam**	**arrivé** trop tard
„	**abamus**	télégraphiez date d'arrivée
„	**abant**	télégraphiez date arrivée probable
„	**abare**	tous les **articles**
„	**abaris**	autres articles
„	**abatur**	**assortiment**
„	**abimur**	télégraphiez assortiment
„	**abis**	prime d'**assurance**
„	**abitis**	télégraphiez prime d'assurance
„	**abo**	y compris l'assurance
„	**amini**	assurance non comprise
„	**amur**	valeur assurée
„	**andam**	valeur à assurer
„	**andi**	télégraphiez valeur à assurer
„	**andior**	„ „ assurée
„	**andius**	assurance maritime
„	**andos**	couvert par assurance
„	**ans**	police d'assurance
„	**antem**	envoyez police d'assurance
„	**antis**	suivant police d'assurance
„	**antium**	**attendu** prochainement
„	**anto**	télégraphiez quand attendu
„	**antur**	vous attendez-vous à (ce que)?
„	**aremus**	on s'attend à
„	**arent**	n'attendons (n'attends) pas
„	**arere**	sous les **auspices** de
„	**ares**	**autant** que possible
„	**aretis**	**authenticité**
„	**aretur**	**autorisation**
„	**ari**	télégraphiez autorisation
„	**arier**	non autorisé/s à
„	**at**	télégraphiez si autorisé/s à
„	**atae**	pas d'**autres**
„	**atarum**	à l'**avance**
„	**atior**	**avarie** grosse
„	**atius**	avarie particulière
„	**atos**	franc d'avaries particulières
„	**atote**	par suite d'avaries
„	**atum**	marchandise avariée
„	**aturae**	marchandise avariée par suite d'emballage défectueux
„	**aturi**	**avec** ou sans
„	**aturos**	à l'**avenir**
„	**averam**	jusqu'à nouvel **avis**
„	**averit**	faute d'avis
„	**avero**	faute d'avis de votre part
„	**avimus**	faute d'avis contraire
„	**avisse**	sans avis contraire
„	**avisti**	sans avis préalable
„	**avit**	à notre (mon) avis
„	**emus**	télégraphiez votre avis concernant
„	**ent**	suivant l'avis général
„	**es**	place **bancable**
„	**etis**	à la **banque**
„	**etur**	de la part de la banque
„	**or**	sur la banque

exul	**abam**	banqueroute frauduleuse
„	**abamus**	sur la **base** de
„	**abant**	sur la même base
„	**abare**	sur cette base
„	**abaris**	sur quelle base?
„	**abatur**	sur la nouvelle base
„	**abimur**	sur l'ancienne base
„	**abis**	par **bateau**
„	**abitis**	télégraphiez nom du bateau
„	**abo**	**bénéfice** brut
„	**amini**	bénéfice net
„	**amur**	avec bénéfice
„	**andam**	sans bénéfice
„	**andi**	pour bénéficier de
„	**andior**	au **besoin**
„	**andius**	au besoin chez
„	**andos**	au besoin chez nous (moi)
„	**ans**	au besoin chez vous
„	**antem**	avons (ai) besoin
„	**antis**	avons (ai) besoin urgent
„	**antium**	au fur et à mesure des besoins
„	**anto**	pas besoin
„	**antur**	besoins sont importants
„	**aremus**	besoins sont insignifiants
„	**arent**	pour les besoins urgents
„	**arere**	pour nos (mes) besoins
„	**ares**	pour vos besoins
„	**aretis**	en **blanc**
„	**aretur**	grièvement **blessé/s**
„	**ari**	légèrement „
„	**arier**	en **bloc**
„	**at**	à **bord**
„	**atae**	tous à bord
„	**atarum**	par dessus bord
„	**atior**	mise à bord
„	**atius**	en **bourse**
„	**atos**	**brevet** No.
„	**atote**	brevet enregistré
„	**atum**	invention pas brevetable
„	**aturae**	le **bruit** court que
„	**aturi**	d'après un bruit qui court
„	**aturos**	il y a des bruits de
„	**averam**	bruits pas confirmés
„	**averit**	**brut** kilo
„	**avero**	brut pour net
„	**avimus**	télégraphiez poids brut
„	**avisse**	**bureau** restant
„	**avisti**	adressez bureau restant
„	**avit**	dans ce **but**
„	**emus**	**ça** et là
„	**ent**	**cahier** des charges
„	**es**	envoyez cahier des charges
„	**etis**	**calculez** à
„	**etur**	„ sur la base de
„	**or**	calculé „ „ „ „

Divers p. 24

fals	**abam**	**campagne** actuelle
„	**abamus**	campagne précédente
„	**abant**	campagne prochaine
„	**abare**	d'ici à fin de campagne
„	**abaris**	la campagne a commencé
„	**abatur**	„ „ est terminée
„	**abimur**	par **canal**
„	**abis**	par canal de Suez
„	**abitis**	le **capitaine** rapporte
„	**abo**	télégraphiez nom du capitaine
„	**amini**	**capital** nominal
„	**amur**	„ social
„	**andam**	„ „ est suffisant
„	**andi**	„ „ est insuffisant
„	**andior**	„ versé
„	**andius**	„ non encore versé
„	**andos**	„ insuffisant
„	**ans**	„ a été augmenté (à)
„	**antem**	„ évalué à
„	**antis**	par suite du manque de capital
„	**antium**	n'ont (n'a) pas de capital
„	**anto**	toute la **cargaison**
„	**antur**	par **carte** postale
„	**aremus**	suivant carte postale
„	**arent**	en **cas** de
„	**arere**	en cas de non-
„	**ares**	dans ce cas
„	**aretis**	dans tous les cas
„	**aretur**	le cas échéant
„	**ari**	suivant **catalogue**
„	**arier**	envoyez catalogue
„	**at**	à **cause** de
„	**atae**	télégraphiez cause de
„	**atarum**	causé par
„	**atior**	contre **caution** (de)
„	**atius**	remettez montant de caution
„	**atos**	remis „ „ „
„	**atote**	pas de caution
„	**atum**	cautionnement provisoire
„	**aturae**	pour **cent**
„	**aturi**	combien de pour cent?
„	**aturos**	il est **certain** (que)
„	**averam**	à peu près certain
„	**averit**	pas encore certain
„	**avero**	envoyez **certificat**
„	**avimus**	„ „ légalisé
„	**avisse**	envoyé „
„	**avisti**	„ „ légalisé
„	**avit**	certificat légalisé
„	**emus**	certificats légalisés
„	**ent**	par suite de la **chaleur**
„	**es**	**Chambre** de Commerce
„	**etis**	il y a des **chances**
„	**etur**	télégraphiez s'il y a des chances
„	**or**	il n'y a pas de chances

foed	**abam**	s'il y a des chances
„	**abamus**	au **change** de
„	**abant**	au change du jour
„	**abare**	au change le plus favorable
„	**abaris**	sur la base du change de
„	**abatur**	télégraphiez change
„	**abimur**	télégraphiez votre avis sur le change
„	**abis**	par suite de dépréciation du change
„	**abitis**	par suite des fluctuations du change
„	**abo**	pas de changement
„	**amini**	„ „ „ dans la situation
„	**amur**	il n'y aura pas de changement
„	**andam**	par suite de changements
„	**andi**	à la **charge** de
„	**andior**	à notre (ma) charge
„	**andius**	à votre charge
„	**andos**	**chargement** complet
„	**ans**	prêt à charger
„	**antem**	**charte-partie**
„	**antis**	par **chemin** de fer
„	**antium**	envoyez **chèque** (sur)
„	**anto**	envoyé chèque (sur)
„	**antur**	chèque No.
„	**aremus**	annulez chèque
„	**arent**	ne payez pas chèque
„	**arere**	chèque a été payé
„	**ares**	„ n'a pas été payé
„	**aretis**	chèque égaré, annulez
„	**aretur**	compte-chèque
„	**ari**	à valoir sur compte-chèque
„	**arier**	en **chiffres** ronds
„	**at**	au **choix** de
„	**atae**	au choix de l'acheteur
„	**atarum**	au choix du vendeur
„	**atior**	notre (mon) choix
„	**atius**	votre choix
„	**atos**	1er choix
„	**atote**	2e choix
„	**atum**	suivant choix
„	**aturae**	par suite de **choléra**
„	**aturi**	par suite du **chômage**
„	**aturos**	suivant les **circonstances**
„	**averam**	dans les circonstances actuelles
„	**averit**	à moins de circonstances exceptionnelles
„	**avero**	d'après **circulaire**
„	**avimus**	envoyez circulaire
„	**avisse**	**circulation**
„	**avisti**	en circulation
„	**avit**	1ère **classe**
„	**emus**	2ième „
„	**ent**	3ième „
„	**es**	billet 1ère classe
„	**etis**	billet 2ième „
„	**etur**	
„	**or**	

form	abam	pour satisfaire et conserver le **client**
„	abamus	de la part de la clientèle
„	abant	jusqu'à la **clôture**
„	abare	avant „ „
„	abaris	après „ „
„	abatur	pas encore clôturé
„	abimur	nouveau **Code**
„	abis	ancien Code
„	abitis	d'après nouveau Code [code
„	abo	envoyez 1 exemplaire de ce
„	amini	envoyez 2 exemplaires de ce
„	amur	par **colis** postal [code
„	andam	expédiez 1 colis postal
„	andi	„ 2 „ postaux
„	andior	„ 3 „ „
„	andius	„ 4 „ „
„	andos	par suite d'une **collision**
„	ans	**combien** de
„	antem	télégraphiez combien de
„	antis	télégraphiez approximativement combien de
„	antium	débordé/s de **commandes**
„	anto	nous sommes (je suis) débordé/s de commandes
„	antur	en cas de commande
„	aremus	au **commencement**
„	arent	jusqu'au commencement
„	arere	dès le commencement
„	ares	à la **commission**
„	aretis	commission en plus
„	aretur	y compris commission
„	ari	„ „ moitié commission
„	arier	commission et ducroire
„	at	télégraphiez commission
„	atae	1/2 % de commission
„	atarum	1 % „ „
„	atior	2 % „ „
„	atius	2 1/2 % „ „
„	atos	3 % „ „
„	atote	4 % „ „
„	atum	5 % „ „
„	aturae	10 % „ „
„	aturi	télégraphiez commission et ducroire
„	aturos	communications interrompues
„	averam	en **comparaison** de
„	averit	**compensation**
„	avero	point de compensation
„	avimus	par compensation
„	avisse	cours de compensation
„	avisti	pas encore **complet**
„	avit	pour completer il faut encore
„	emus	complètement
„	ent	presque complètement
„	es	**composé** de
„	etis	composé principalement de
„	etur	non **compris** (dans)
„	or	télégraphiez si compris (dans)

foss	abam	n'avons (n'ai) pas compris
„	abamus	avons (ai) compris comme suit . . . si correct, pas nécessaire répondre
„	abant	net **comptant**
„	abare	comptant 2 % d'escompte
„	abaris	„ 2 1/2 „ „
„	abatur	„ 3 „ „
„	abimur	ne traitez qu'au comptant
„	abis	**compte** courant
„	abitis	notre (mon) compte
„	abo	votre compte
„	amini	leur (son) compte
„	amur	pour notre compte
„	andam	„ mon compte
„	andi	„ votre compte
„	andior	„ leur (son) compte
„	andius	à valoir sur compte
„	andos	compte Profits et Pertes
„	ans	„ à demi
„	antem	pour solde de compte
„	antis	compte nouveau
„	antium	„ d'achat
„	anto	„ de vente
„	antur	à compte nouveau
„	aremus	envoyez compte fictif
„	arent	à compter de
„	arere	vous pouvez compter sur
„	ares	télégraphiez si vous comptez
„	aretis	pouvons-nous compter sur?
„	aretur	à une **concession**
„	ari	point de concession
„	arier	par suite d'un **concordat**
„	at	moyennant un concordat
„	atae	approuvons (approuve) proposition de concordat
„	atarum	homologation du concordat
„	atior	concordataire(s)
„	atius	assurez **concours** de
„	atos	impossible de concourir
„	atote	jusqu'à **concurrence** de
„	atum	de la part de la concurrence
„	aturae	vu la forte concurrence
„	aturi	télégraphiez prix de la concurrence
„	aturos	pour faire face à la concurrence
„	averam	la concurrence est énorme
„	averit	télégraphiez **conditions**
„	avero	à quelles conditions?
„	avimus	aux conditions de
„	avisse	„ anciennes conditions
„	avisti	„ nouvelles conditions
„	avit	„ conditions connues
„	emus	„ „ stipulées
„	ent	mêmes conditions (que)
„	es	mêmes conditions que fois passée
„	etis	nous acceptons (j'accepte) conditions mentionnées
„	etur	télégraphiez si vous acceptez conditions mentionnées
„	or	télégraphiez si l'on accepte conditions mentionnées

fren	abam	conditions inacceptables
„	abamus	à ces conditions
„	abant	à conditions favorables
„	abare	„ „ très favorables
„	abaris	„ „ plus favorables
„	abatur	„ „ moins favorables
„	abimur	conditions habituelles
„	abis	conditionnement
„	abitis	suivant conditionnement
„	abo	sincères condoléances
„	amini	ne mérite(nt) point de **confiance**
„	amur	„ „ que peu de confiance
„	andam	mérite(nt) pleine confiance
„	andi	vu le manque de confiance
„	andior	**confidentiel**(lement)
„	andius	ceci à titre confidentiel
„	andos	**confirmation**
„	ans	télégraphiez confirmation
„	antem	sauf votre confirmation
„	antis	sauf notre (ma) confirmation
„	antium	pas **conforme**(s) à
„	anto	en conformité (de)
„	antur	en **congé**
„	aremus	**connaissement**
„	arent	connaissements
„	arere	envoyez connaissement
„	ares	„ connaissements
„	aretis	série complète de connaisse- [ments
„	aretur	connaissement daté
„	ari	contre remise du connaissement
„	arier	contre remise du connaissement et de la facture
„	at	connaissement à ordre
„	atae	connaissement, facture et police d'assurance
„	atarum	sur connaissement
„	atior	pas reçu connaissement
„	atius	ne **connaissons** (ne connais) [pas
„	atos	encore inconnu
„	atote	sauf **consentement** (de)
„	atum	sauf votre consentement
„	aturae	sauf notre (mon) consentement
„	aturi	par **conséquent**
„	aturos	en **consignation**
„	averam	à la consignation
„	averit	à valoir sur consignation
„	avero	vendez en consignation
„	avimus	vendu „ „
„	avisse	vendu en consignation et [réalisé
„	avisti	**consommateur(s)**
„	avit	de la part des consommateurs
„	emus	à la consommation
„	ent	consommation locale
„	es	consommation intérieure
„	etis	pour la consommation
„	etur	„ „ „ locale
„	or	**Consul** Allemand

furi	abam	**Consul** Américain
„	abamus	„ Anglais
„	abant	„ Français
„	abare	Consulat Allemand
„	abaris	„ Américain
„	abatur	„ Anglais
„	abimur	„ Français
„	abis	**contestation**
„	abitis	en cas de contestation
„	abo	par **continuation**
„	amini	par suite de continuation (de)
„	amur	sauf indications **contraires**
„	andam	contrairement (à)
„	andi	envoyez **contrat**
„	andior	suivant contrat
„	andius	contrat signé
„	andos	nouveau contrat
„	ans	ancien contrat
„	antem	contrat en cours
„	antis	contrats en cours
„	antium	à valoir sur contrat
„	anto	à valoir sur contrat en cours
„	antur	à l'expiration du contrat
„	aremus	contrat renouvelable
„	arent	contrat renouvelé
„	arere	résilions contrat
„	ares	contrat non encore exécuté
„	aretis	sous contrat
„	aretur	par **contre**
„	ari	contremarque
„	arier	contremarques
„	at	contre-ordre
„	atae	sauf contre-ordre
„	atarum	sans contre-partie
„	atior	par suite de contrevention
„	atius	sous **contrôle**
„	atos	en cas de **convenance**
„	atote	si celà vous convient ainsi
„	atum	télégraphiez si convient
„	aturae	ne convient pas
„	aturi	il semble convenable
„	aturos	ne semble pas convenable
„	averam	comme convenu
„	averit	suivant **convention**
„	avero	par suite d'une convention
„	avimus	**coopération**
„	avisse	avec coopération de
„	avisti	envoyez **copie**
„	avit	envoyé copie
„	emus	pas reçu copie
„	ent	égaré, envoyez copie
„	es	envoyez copie de toute la correspondance
„	etis	copie de facture
„	etur	envoyez copie de facture
„	or	**correspondance**

fusc	**abam**	de ce **coté**
„	**abamus**	de notre (mon) coté
„	**abant**	de votre coté
„	**abare**	télégraphiez cote
„	**abaris**	„ cotes
„	**abatur**	cotes nominales faute d'acheteurs
„	**abimur**	il y a du **coulage**
„	**abis**	poids diminué par coulage
„	**abitis**	ex-**coupon**
„	**abo**	coupon d'à-compte
„	**amini**	„ semestriel
„	**amur**	„ trimestriel
„	**andam**	à l'échéance du coupon
„	**andi**	pour paiement de coupon
„	**andior**	feuille de coupon
„	**andius**	au **courant** [(de)
„	**andos**	tenez-nous (moi) au courant
„	**ans**	dans le courant de
„	**antem**	par premier **courrier**
„	**antis**	courrier pas arrivé
„	**antium**	1/8 % de **courtage**
„	**anto**	1/4 % „ „
„	**antur**	1/2 % „ „
„	**aremus**	3/4 % „ „
„	**arent**	1 % „ „
„	**arere**	estimons (estime) **coût** à
„	**ares**	coût estimé à
„	**aretis**	pour couvrir le coût
„	**aretur**	couvrant à peine le coût
„	**ari**	coûtera à peu près [pas
„	**arier**	pourvu que coût ne dépasse
„	**at**	coût et frêt
„	**atae**	„ frêt et assurance
„	**atarum**	coûte que coûte
„	**atior**	comme de coutume
„	**atius**	moins que de coutume
„	**atos**	plus „ „ „
„	**atote**	en **couverture** (de)
„	**atum**	contre couverture (en)
„	**aturae**	sans couverture
„	**aturi**	en couverture partielle
„	**aturos**	„ „ totale
„	**averam**	au couvert à l'étendue de
„	**averit**	on **craint** (que)
„	**avero**	il est à craindre (que)
„	**avimus**	de la part des **créanciers**
„	**avisse**	solde de **crédit**
„	**avisti**	à valoir sur solde à votre crédit
„	**avit**	à valoir sur solde à notre (mon)
„	**emus**	à crédit [crédit
„	**ent**	votre crédit
„	**es**	notre (mon) crédit
„	**etis**	pas de crédit
„	**etur**	télégraphiez si l'on pourrait accorder crédit et à quelle étendue
„	**or**	il est préférable de n'accorder aucun crédit

gale	**abam**	petits découverts renouvelables après couverture
„	**abamus**	petits découverts renouvelables après couverture semblent admissibles
„	**abant**	les négociants consultés accordent petits découverts renouvelables après couverture
„	**abare**	à valoir sur crédit
„	**abaris**	à notre (mon) crédit
„	**abatur**	à votre crédit
„	**abimur**	à leur (son) crédit
„	**abis**	crédit épuisé, renouvelez
„	**abitis**	„ télégraphique
„	**abo**	„ renouvelé
„	**amini**	renouvelez crédit
„	**amur**	crédit illimité
„	**andam**	„ modéré
„	**andi**	„ restreint
„	**andior**	mérite(nt) crédit blanc de
„	**andius**	ne mérite(nt) point de crédit
„	**andos**	une **crise** est imminente
„	**ans**	la crise a passé
„	**antem**	crise pas encore passée
„	**antis**	nous **croyons** (je crois) que
„	**antium**	il y a lieu de croire que
„	**anto**	on ne croit pas (que)
„	**antur**	par suite d'un **cyclone**
„	**aremus**	en **danger**
„	**arent**	hors de danger
„	**arere**	télégraphiez s'il y a danger
„	**ares**	pas de danger
„	**aretis**	télégraphiez **date**
„	**aretur**	télégraphiez date approximative
„	**ari**	„ „ d'arrivée
„	**arier**	„ „ de départ
„	**at**	date d'annulation
„	**atae**	8 jours de date
„	**atarum**	15 „ „ „
„	**atior**	30 „ „ „
„	**atius**	45 „ „ „
„	**atos**	60 „ „ „
„	**atote**	3 mois „ „
„	**atum**	4 „ „ „
„	**aturae**	6 „ „ „
„	**aturi**	encore **davantage**
„	**aturos**	pas davantage
„	**averam**	par suite de **débordements**
„	**averit**	télégraphiez **débours**
„	**avero**	au **début**
„	**avimus**	dès le début
„	**avisse**	**déclaration**
„	**avisti**	à la déclaration
„	**avit**	déclaration inexacte
„	**emus**	
„	**ent**	
„	**es**	
„	**etis**	
„	**etur**	
„	**or**	

gel	**abam**	par suite de déclaration inexacte
„	**abamus**	par suite de déclaration de valeur inférieure
„	**abant**	valeur déclarée
„	**abare**	télégraphiez valeur déclarée
„	**abaris**	valeur à déclarer
„	**abatur**	télégraphiez valeur à déclarer
„	**abimur**	envoyez **décompte**
„	**abis**	à **découvert**
„	**abitis**	découvert est important
„	**abo**	notre (mon) découvert est considérable
„	**amini**	**dédouanement**
„	**amur**	frais de dédouanement
„	**andam**	dédouanement à la frontière
„	**andi**	sans **déduction**
„	**andior**	sous déduction des frais
„	**andius**	pour nous (m') en **défaire**
„	**andos**	à **défaut**
„	**ans**	font (fait) totalement défaut
„	**antem**	il y a un **déficit** de
„	**antis**	il y a un déficit sensible
„	**antium**	par suite de déficit
„	**anto**	„ „ „ „ sensible
„	**antur**	pour faire face au déficit
„	**aremus**	**dègâts** considérables
„	**arent**	dégâts pas sensibles
„	**arere**	télégraphiez étendue des dégâts
„	**ares**	dégâts matériels évalués à
„	**aretis**	par suite de dégâts
„	**aretur**	en **dehors** de
„	**ari**	sans aucun **délai**
„	**arier**	dernier délai
„	**at**	par suite de délai
„	**atae**	à **délester** à
„	**atarum**	les **demandes** sont nombreuses
„	**atior**	sur demande
„	**atius**	demande locale
„	**atos**	demande de l'intérieur
„	**atote**	au **départ**
„	**atum**	**dépend** entièrement de
„	**aturae**	condamnés aux **dépens**
„	**aturi**	en **dépôt**
„	**aturos**	il faut un dépôt de
„	**averam**	certificat de dépôt
„	**averit**	à valoir sur dépôt
„	**avero**	au **dépourvu**
„	**avimus**	étant dépourvu(s)
„	**avisse**	complètement dépourvu(s)
„	**avisti**	**dépréciations**
„	**avit**	par suite de dépréciation
„	**emus**	**depuis** hier
„	**ent**	**dès** ce jour
„	**es**	comme **désiré**
„	**etis**	au-**dessous** de
„	**etur**	pas au-dessous de
„	**or**	au-**dessus** de
gest	**abam**	pas au-dessus de
„	**abamus**	à **destination**
„	**abant**	destination directe
„	**abare**	„ inconnue
„	**abaris**	télégraphiez destination
„	**abatur**	„ **détails**
„	**abimur**	détails suivent par lettre
„	**abis**	en détail
„	**abitis**	marchand(s) en détail
„	**abo**	envoyez détail
„	**amini**	les détails manquent
„	**amur**	**détenteurs** réservés
„	**andam**	détenteurs empressés
„	**andi**	à **déterminer** d'avance
„	**andior**	par suite de **détournements**
„	**andius**	au **détriment** de
„	**andos**	à notre (mon) détriment
„	**ans**	à votre détriment
„	**antem**	**dette** flottante
„	**antis**	**développement**
„	**antium**	**différence** en plus
„	**anto**	différence en moins
„	**antur**	il est **difficile** de
„	**aremus**	très difficile
„	**arent**	par suite de difficultés
„	**arere**	télégraphiez **dimensions**
„	**ares**	par suite de **diminution**
„	**aretis**	sous la **direction** de
„	**aretur**	nous avons (j'ai) **disponible**
„	**ari**	nous n'avons (je n'ai) rien disponible
„	**arier**	pas disponible
„	**at**	n'est plus disponible
„	**atae**	télégraphiez si disponible
„	**atarum**	non encore disponible
„	**atior**	le disponible et livrable
„	**atius**	télégraphiez combien disponible
„	**atos**	télégraphiez **dispositions**
„	**atote**	suivez dispositions de
„	**atum**	à la disposition de
„	**aturae**	à votre disposition
„	**aturi**	„ notre (ma) „
„	**aturos**	„ leur (sa) „
„	**averam**	la marchandise se trouve à votre disposition
„	**averit**	on **dit** que
„	**avero**	d'après ce qu'on dit
„	**avimus**	**dividende** probablement
„	**avisse**	point de dividende
„	**avisti**	ex-dividende
„	**avit**	dividende provisoire
„	**emus**	à compte sur dividende
„	**ent**	dans le **dock**
„	**es**	contre **documents**
„	**etis**	„ „ d'expédition
„	**etur**	livraison contre documents d'expédition
„	**or**	envoyez documents

glob	**abam**	franco **domicile**
„	**abamus**	payable au domicile de
„	**abant**	domicilié payable chez
„	**abare**	télégraphiez étendue du **dommage**
„	**abaris**	par suite de dommage
„	**abatur**	par suite de graves dommages
„	**abimur**	dommages-intérêts
„	**abis**	à la **douane**
„	**abitis**	tâchez de **doubler** quantité
„	**abo**	**droits** d'entrée
„	**amini**	„ „ compris
„	**amur**	„ „ non compris
„	**andam**	„ d'exportation
„	**andi**	franco de droits d'entrée
„	**andior**	droits d'entrée augmentés à
„	**andius**	„ „ réduits à
„	**andos**	droits d'entrée seront probablement augmentés à
„	**ans**	droits d'entrée seront problement diminués à
„	**antem**	exempt(s) de droits d'entrée
„	**antis**	droits ad valorem [d'entrée
„	**antium**	télégraphiez taux de droits
„	**anto**	pour éviter de payer droits d'entrée
„	**antur**	nouveau tarif douanier entrera en vigueur
„	**aremus**	pour éviter de payer les nouveaux droits d'entrée il faut expédier avant
„	**arent**	en **droiture**
„	**arere**	**dû** à
„	**ares**	dû, pas encore arrivé
„	**aretis**	**ducroire** 2%
„	**aretur**	„ 2½%
„	**ari**	„ 3%
„	**arier**	„ 4%
„	**at**	„ 5%
„	**atae**	télégraphiez taux de ducroire
„	**atarum**	envoyez **duplicata**
„	**atior**	égaré, envoyez duplicata
„	**atius**	télégraphiez **durée**
„	**atos**	voici durée probable
„	**atote**	probablement de courte durée
„	**atum**	„ „ longue „
„	**aturae**	plein **d'eau**
„	**aturi**	par suite du manque d'eau
„	**aturos**	à un **écart** de
„	**averam**	**échantillon**
„	**averit**	échantillons
„	**avero**	suivant échantillon
„	**avimus**	„ échantillons
„	**avisse**	„ votre échantillon
„	**avisti**	„ vos échantillons
„	**avit**	„ notre (mon) échantillon
„	**emus**	„ nos (mes) échantillons
„	**ent**	„ échantillon(s) de qualité
„	**es**	pas reçu échantillon(s)
„	**etis**	envoyez échantillon
„	**etur**	„ échantillons
„	**or**	„ échantillons pris au hazard

grav	**abam**	envoyez échantillon en cotant prix de
„	**abamus**	„ échantillons „ „ „ „
„	**abant**	envoyez contre-échantillon
„	**abare**	contre-échantillon
„	**abaris**	„ -échantillons
„	**abatur**	envoyé échantillon
„	**abimur**	envoyé échantillons
„	**abis**	envoyez collection d'échantillons
„	**abitis**	envoyé „ „ [tillon
„	**abo**	marchandise pas conforme à l'échantillon
„	**amini**	marchandise pas conforme aux échantillons, envoi refusé
„	**amur**	impossible de conclure faute d'échantillons
„	**andam**	à **l'échéance**
„	**andi**	télégraphiez échéance
„	**andior**	à longue échéance
„	**andius**	à courte échéance
„	**andos**	à échéance plus éloignée
„	**ans**	à plus brève échéance
„	**antem**	sur une vaste **échelle**
„	**antis**	bateau s'est **échoué** (à)
„	**antium**	pour **économiser**
„	**anto**	pour **écouler**
„	**antur**	„ „ le prix est réduit
„	**aremus**	1ère **édition**
„	**arent**	2ème „
„	**arere**	**effectué** par
„	**ares**	à effectuer (par)
„	**aretis**	à cet **effet**
„	**aretur**	en effet
„	**ari**	égal à
„	**arier**	à peu près **égal** à
„	**at**	à cet **égard**
„	**atae**	à leur (son) égard
„	**atarum**	**emballage** compris
„	**atior**	emballage non-compris
„	**atius**	veillez à ce que l'emballage soit irréprochable
„	**atos**	par suite d'emballage défectueux
„	**atote**	un **embargo** a été mis (sur)
„	**atum**	par suite d'embargo
„	**aturae**	embargo levé
„	**aturi**	prompt **embarquement**
„	**aturos**	embarquement immédiat
„	**averam**	„ éloigné
„	**averit**	„ rapproché
„	**avero**	**empêché** par
„	**avimus**	nous n'en avons (je n'en ai) pas l'**emploi**
„	**avisse**	remettez aussitôt **encaissé**
„	**avisti**	à l'encaissement
„	**avit**	présenté à l'encaissement
„	**emus**	à valoir sur encaissement
„	**ent**	aux **enchères**
„	**es**	„ „ de ce jour
„	**etis**	„ dernières enchères
„	**etur**	„ prochaines enchères
„	**or**	

décl.- p. 30

greg	abam	prochaines enchères auront
„	abamus	vendu aux enchères [lieu
„	abant	acheté „ „
„	abare	**encombrement**
„	abaris	pas **encore**
„	abatur	si encore
„	abimur	si pas encore
„	abis	faute **d'endossement**
„	abitis	endossé par
„	abo	sans **engagement**
„	amini	nous nous engageons (je m'engage) à
„	amur	engagements
„	andam	„ sont importants
„	andi	**entendez**-vous avec
„	andior	à défaut d'entente
„	andius	par suite d'entente
„	andos	en **entier**
„	ans	**entr**'autres
„	antem	par l'**entremise** de
„	antis	en **entrepôt**
„	antium	mise à l'entrepôt
„	anto	aux **environs** de
„	antur	d'**envoi** de
„	aremus	avez-vous reçu envoi?
„	arent	envoi pas encore arrivé
„	arere	aussitôt envoi arrivé télégra-
„	ares	télégraphiez **épaisseur** [phiez
„	aretis	toutes les **époques**
„	aretur	complètement **épuisé**
„	ari	presque complètement épuisé
„	arier	capitaine et **équipage**
„	at	équipage sauvé
„	atae	à l'**équivalent** de
„	atarum	par erreur
„	atior	par suite d'une erreur
„	atius	erreur rectifiée
„	atos	il doit y avoir erreur, examinez et rectifiez s'il y a lieu
„	atote	sans **escale**
„	atum	avec escale à
„	aturae	net sans **escompte**
„	aturi	télégraphiez taux d'escompte
„	aturos	1% d'escompte
„	averam	$1\frac{1}{2}$% „
„	averit	2% „
„	avero	$2\frac{1}{2}$% „
„	avimus	3% „
„	avisse	4% „
„	avisti	5% „
„	avit	en **espèces**
„	emus	on **espère**
„	ent	il y a peu d'espoir
„	es	il n'y a point d'espoir
„	etis	**essai** a pleinement réussi
„	etur	essai n'a pas réussi
„	or	ordre d'essai

gyps	abam	à titre d'essai
„	abamus	non **estampillé**
„	abant	muni de l'estampille
„	abare	télégraphiez **estimation**
„	abaris	au-dessous de l'estimation
„	abatur	au-dessus „ „
„	abimur	nous estimons (j'estime) à
„	abis	on estime généralement que
„	abitis	suivant l'estimation générale
„	abo	et/ou
„	amini	s'est **établi** à son compte
„	amur	va s'établir à son compte
„	andam	en bon **état**
„	andi	en mauvais état
„	andior	en parfait état
„	andius	en très mauvais état
„	andos	état laisse à désirer
„	ans	arrivé en bon état
„	antem	télégraphiez état de
„	antis	à l'**étendue** de
„	antium	télégraphiez étendue de
„	anto	à l'étendue indiquée
„	antur	à l'étendue stipulée
„	aremus	à l'**étranger**
„	arent	de l'étranger
„	arere	**évalué** à
„	ares	évalué à environ
„	aretis	dans l'**éventualité** de
„	aretur	il est **évident** que
„	ari	évitez autant que possible
„	arier	**examinez** de nouveau
„	at	à titre d'**exception**
„	atae	à l'exception de
„	atarum	non **exclus**
„	atior	envoyez 1 **exemplaire** (de)
„	atius	envoyez 2 exemplaires (de)
„	atos	exempt de
„	atote	garanti **exempt** de
„	atum	pour l'**exercice** en cours
„	aturae	n'**existe**(nt) pas
„	aturi	il n'existe pas de
„	aturos	**expédiez** aussitôt que possible
„	averam	expédierons (expédierai) aussitôt que possible
„	averit	expédiez autant que possible
„	avero	par **expert**
„	avimus	par experts
„	avisse	désignez un expert
„	avisti	moyennant expertise
„	avit	envoyez à l'expertise
„	emus	frais d'expertise
„	ent	**explications** par lettre
„	es	télégraphiez explication
„	etis	**exploitation**
„	etur	**exportation**
„	or	pour exportation

gyr	**abam**	par **express**
„	**abamus**	expressément
„	**abant**	de la part des fabricants
„	**abare**	en cours de fabrication
„	**abaris**	pris en fabrique
„	**abatur**	de quelle **façon?**
„	**abimur**	envoyez **facture**
„	**abis**	„ „ fictive
„	**abitis**	suivant votre facture
„	**abo**	suivant notre (ma) facture
„	**amini**	factures légalisées
„	**amur**	envoyez factures légalisées
„	**andam**	facture en double
„	**andi**	montant de facture
„	**andior**	pas reçu facture
„	**andius**	avec **faculté** de
„	**andos**	tâchez d'obtenir faculté de
„	**ans**	fait **faillite**
„	**antem**	par suite de faillite
„	**antis**	fait faillite une fois
„	**antium**	„ „ plusieurs fois
„	**anto**	compromis dans la faillite
„	**antur**	télégraphiez si compromis dans la faillite
„	**aremus**	pas compromis dans la faillite
„	**arent**	faillite est imminente
„	**arere**	il **faut**
„	**ares**	il faut absolument
„	**aretis**	s'il faut
„	**aretur**	télégraphiez s'il faut
„	**ari**	il ne faut pas
„	**arier**	**faute** de
„	**at**	faute de quoi
„	**atae**	en votre **faveur**
„	**atarum**	en notre faveur
„	**atior**	en ma faveur
„	**atius**	en leur (sa) faveur
„	**atos**	**félicitations**
„	**atote**	par suite des **fêtes**
„	**atum**	avant les fêtes
„	**aturae**	après les fêtes
„	**aturi**	**filon** a été coupé
„	**aturos**	jusqu'à la **fin** de
„	**averam**	avant fin de
„	**averit**	d'ici à fin de
„	**avero**	pour mettre fin à
„	**avimus**	**flottant** à destination de
„	**avisse**	en une seule **fois**
„	**avisti**	1 fois
„	**avit**	2 fois
„	**emus**	3 fois
„	**ent**	la prochaine fois
„	**es**	la dernière fois
„	**etis**	mal fondé
„	**etur**	faute de **fonds**
„	**or**	avons (ai) besoin de fonds

hal	**abam**	assez de fonds
„	**abamus**	cas de **force** majeure
„	**abant**	à **forfait**
„	**abare**	télégraphiez taux à forfait
„	**abaris**	nouvelle **forme**
„	**abatur**	ancienne forme
„	**abimur**	sans **frais**
„	**abis**	à leurs (ses) frais
„	**abitis**	à nos (mes) frais
„	**abo**	à vos frais
„	**amini**	net de tous frais
„	**amur**	télégraphiez frais
„	**andam**	aux frais, risques et périls
„	**andi**	à vos frais, risques et périls
„	**andior**	à nos (mes) frais, risques et périls
„	**andius**	les frais sont trop élevés
„	**andos**	aux frais les plus réduits
„	**ans**	au **frêt** le plus réduit
„	**antem**	télégraphiez taux de frêt
„	**antis**	frêt d'occasion
„	**antium**	nous pourrons (je pourrai) probablement obtenir frêt d'occasion
„	**anto**	par suite de frêt d'occasion
„	**antur**	faute de frêt
„	**aremus**	par suite du **froid**
„	**arent**	à la **frontière**
„	**arere**	jusqu'à la frontière
„	**ares**	au **fur** et à mesure (de)
„	**aretis**	en **fûts**
„	**aretur**	fûts perdus
„	**ari**	avec **garantie** (de)
„	**arier**	„ „ usuelle
„	**at**	sans garantie
„	**atae**	qualité garantie
„	**atarum**	en **gare** d'arrivée
„	**atior**	en gare de départ
„	**atius**	par suite de **gelée**
„	**atos**	en **général**
„	**atote**	pour notre (ma) **gouverne**
„	**atum**	pour votre gouverne
„	**aturae**	en **grève**
„	**aturi**	par suite de grève
„	**aturos**	grève est imminente
„	**averam**	grève est terminée
„	**averit**	on a proclamé la grève générale
„	**avero**	comme d'**habitude**
„	**avimus**	habituellement
„	**avisse**	sous **hangar**
„	**avisti**	de la part des **haussiers**
„	**avit**	en dernière **heure**
„	**emus**	à l'heure actuelle
„	**ent**	de bonne heure
„	**es**	à quelle heure?
„	**etis**	par suite d'**humidité**
„	**etur**	d'**ici** à
„	**or**	d'ici là

hiem	**abam**	d'ici quelque temps
„	**abamus**	on **ignore** si
„	**abant**	nous ignorons (j'ignore) si
„	**abare**	**immédiatement**
„	**abaris**	d'aucune **importance**
„	**abatur**	vu l'importance de l'affaire
„	**abimur**	c'est très important
„	**abis**	pour l'**importation**
„	**abitis**	n'**importe** quel
„	**abo**	il est **impossible**
„	**amini**	vu l'impossibilité (de)
„	**amur**	impossible faute de
„	**andam**	impossible par suite de
„	**andi**	si impossible
„	**andior**	sauf **imprévu**
„	**andius**	par suite d'**incendie**
„	**andos**	un incendie s'est déclaré
„	**ans**	encore **incertain**
„	**antem**	action en **indemnité**
„	**antis**	**indépendemment** (de)
„	**antium**	sauf **indications** contraires
„	**anto**	comme indiqué
„	**antur**	**indispensable(s)**
„	**aremus**	il est indispensable de
„	**arent**	si indispensable (de)
„	**arere**	il est **inexact** (que)
„	**ares**	en première **instance**
„	**aretis**	**instructions**
„	**aretur**	télégraphiez instructions
„	**ari**	suivant instructions (de)
„	**arier**	„ vos instructions
„	**at**	„ nos(mes)instructions
„	**atae**	suivez instructions de
„	**atarum**	attendez instructions ultérieures
„	**atior**	sauf instructions contraires
„	**atius**	à notre (mon) **insu**
„	**atos**	par suite d'**insuffisance** (de)
„	**atote**	insuffisant|e|s
„	**atum**	insuffisamment
„	**aturae**	**intentez** une action (contre)
„	**aturi**	intenté „ „ „
„	**aturos**	intenterons (intenterai) une action (contre)
„	**averam**	avez-vous l'**intention** de?
„	**averit**	avons (ai) l'intention (de)
„	**avero**	n'avons (n'ai) pas intention (de)
„	**avimus**	la partie **intéressée**
„	**avisse**	les parties intéressées
„	**avisti**	informez parties intéressées
„	**avit**	avons (ai) informé parties intéressées
„	**emus**	dans leurs (ses) intérêts
„	**ent**	au mieux de nos intérêts
„	**es**	„ „ „ vos „
„	**etis**	agissez au mieux de nos (mes) intérêts
„	**etur**	agirons (agirai) au mieux de vos intérêts
„	**or**	perte d'intérêts

hum	**abam**	intérêts annuels
„	**abamus**	intérêts de retard
„	**abant**	télégraphiez taux d'intérêt
„	**abare**	$1^{1}/_{2}$ % intérêt annuel
„	**abaris**	2 % „ „
„	**abatur**	$2^{1}/_{2}$ % „ „
„	**abimur**	3 % „ „
„	**abis**	$3^{1}/_{2}$ % „ „
„	**abitis**	4 % „ „
„	**abo**	$4^{1}/_{2}$ % „ „
„	**amini**	5 % „ „
„	**amur**	$5^{1}/_{2}$ % „ „
„	**andam**	6 % „ „
„	**andi**	7 % „ „
„	**andior**	10 % „ „
„	**andius**	payement d'intérêts
„	**andos**	à l'**intérieur**
„	**ans**	de „
„	**antem**	**interruption**
„	**antis**	sans interruption
„	**antium**	par suite d'interruption
„	**anto**	dans l'**intervalle**
„	**antur**	avec intervalle de
„	**aremus**	**intervention**
„	**arent**	**introuvable(s)**
„	**arere**	**irrécouvrable(s)**
„	**ares**	**jeté** par dessus bord
„	**aretis**	par **jour**
„	**aretur**	de jour en jour
„	**ari**	tous les jours
„	**arier**	jours ouvrables
„	**at**	jours fériés
„	**atae**	d'un jour à l'autre
„	**atarum**	dès ce jour
„	**atior**	jusqu'à ce jour
„	**atius**	dans quelques jours
„	**atos**	jour et nuit
„	**atote**	30 jours, mois de livraison non compris
„	**atum**	toute la journée
„	**aturae**	frais **judiciaires**
„	**aturi**	à en **juger** par
„	**aturos**	**jusqu'à**
„	**averam**	jusqu'où
„	**averit**	jusqu'à quand?
„	**avero**	pièces **justificatives**
„	**avimus**	envoyez pièces justificatives
„	**avisse**	de **là**
„	**avisti**	**laissez**-suivre
„	**avit**	contre laissez-suivre
„	**emus**	télégraphiez largeur
„	**ent**	**légalisé** par Consul
„	**es**	„ „ notaire
„	**etis**	légalisation
„	**etur**	légalisation est obligatoire
„	**or**	le lendemain de

imit	**abam**	par **lettre**
„	**abamus**	suivant lettre
„	**abant**	suivant votre lettre
„	**abare**	suivant notre (ma) lettre
„	**abaris**	lettre confidentielle
„	**abatur**	suivant votre lettre confidentielle
„	**abimur**	„ notre(ma) lettre confidentielle
„	**abis**	lettre de commande
„	**abitis**	„ „ voiture
„	**abo**	suivant lettre de commande
„	**amini**	„ „ „ voiture
„	**amur**	envoyez lettre
„	**andam**	envoyez lettres
„	**andi**	envoyé lettre
„	**andior**	lettre de recommandation
„	**andius**	envoyez lettre de recommandation
„	**andos**	détails par lettre
„	**ans**	détails plus complets par lettre
„	**antem**	„ „ „ „ „ de ce
„	**antis**	attendez lettre [jour
„	**antium**	dès réception de lettre
„	**anto**	avant réception de lettre
„	**antur**	reçu aucune lettre
„	**aremus**	réexpédiez lettre
„	**arent**	aura (auront) **lieu**
„	**arere**	quand aura (auront) lieu?
„	**ares**	n'aura (n'auront) pas lieu
„	**aretis**	il y a lieu de
„	**aretur**	sur le lieu
„	**ari**	télégraphiez **limite** (pour)
„	**arier**	„ limites „
„	**at**	au dessus de limite(s)
„	**atae**	„ dessous „ „
„	**atarum**	renouvelez limite(s)
„	**atior**	limite(s) renouvelée(s) à
„	**atius**	réduisez limite(s) (à)
„	**atos**	augmentez limite(s) (à)
„	**atote**	impossible conclure à votre limite
„	**atum**	„ „ „ vos limites
„	**aturae**	ne dépassez pas limite(s)
„	**aturi**	annulez toutes les limites
„	**aturos**	annulez limite
„	**averam**	limite en vigueur jusqu'à révocation
„	**averit**	**liquidation**
„	**avero**	en liquidation
„	**avimus**	par suite de liquidation(s)
„	**avisse**	envoyez **liste**
„	**avisti**	en litige
„	**avit**	quantité **livrée**
„	**emus**	quantité à livrer
„	**ent**	télégraphiez quantité à livrer
„	**es**	contre livraison
„	**etis**	lors de livraison
„	**etur**	à valoir sur quantité à livrer
„	**or**	télégraphiez délai livraison

impi	**abam**	livraison 1	semaine/s dès réception de la commande
„	**abamus**	„ 2	
„	**abant**	„ 3	
„	**abare**	„ 4	
„	**abaris**	„ 5	
„	**abatur**	„ 6	
„	**abimur**	„ 7	
„	**abis**	„ 8	
„	**abitis**	télégraphiez **longueur**	
„	**abo**	**lors** de	
„	**amini**	en 1 **lot**	
„	**amur**	en 2 lots	
„	**andam**	„ lots de	
„	**andi**	„ un seul ou plusieurs lots	
„	**andior**	en **magasin**	
„	**andius**	frais de magasin	
„	**andos**	faites emmagasiner marchandise	
„	**ans**	marchandise se trouve en magasin	
„	**antem**	de première **main**	
„	**antis**	de seconde main	
„	**antium**	en mains (de)	
„	**anto**	main d'oeuvre	
„	**antur**	dès **maintenant**	
„	**aremus**	**malgré** celà	
„	**arent**	par prochaine **malle**	
„	**arere**	**mandat-poste**	
„	**ares**	remettez par mandat-poste	
„	**aretis**	remis „ „	
„	**aretur**	de cette **manière**	
„	**ari**	de la même manière [cédemment	
„	**arier**	de la même manière que pré-	
„	**at**	de la manière habituelle	
„	**atae**	„ „ „ désirée	
„	**atarum**	„ „ „ stipulée	
„	**atior**	„ „ „ mentionnée	
„	**atius**	de quelle manière?	
„	**atos**	de quelque manière que ce soit	
„	**atote**	suivant **manifeste**	
„	**atum**	envoyez manifeste	
„	**aturae**	serment de manifestation	
„	**aturi**	il y a un **manque** de	
„	**aturos**	par suite du manque de	
„	**averam**	manque(nt) complètement	
„	**averit**	**marchandise**(s)	
„	**avero**	pas reçu marchandise	
„	**avimus**	avez-vous reçu marchandise?	
„	**avisse**	bon **marché**	
„	**avisti**	aussi bon marché que possible	
„	**avit**	meilleur marché	
„	**emus**	exceptionnellement bon marché	
„	**ent**	marché animé	
„	**es**	marché très animé	
„	**etis**	marché sans animation	
„	**etur**	„ trop approvisionné	
„	**or**	„ très avantageux	

infl	**abam**	marché en baisse
„	**abamus**	„ calme
„	**abant**	„ „ et sans variation
„	**abare**	„ très calme
„	**abaris**	„ défavorable
„	**abatur**	„ délaissé
„	**abimur**	„ faible
„	**abis**	„ plus faible
„	**abitis**	„ très faible
„	**abo**	„ très ferme
„	**amini**	„ plus ferme
„	**amur**	„ moins ferme
„	**andam**	„ en hausse
„	**andi**	„ indécis
„	**andior**	„ irrité
„	**andius**	„ lourd
„	**andos**	„ tendance fléchissante
„	**ans**	„ mouvementé
„	**antem**	„ nerveux
„	**antis**	„ soutenu
„	**antium**	„ sans variation
„	**anto**	sur le marché
„	**antur**	télégraphiez vos prévisions sur le [marché
„	**aremus**	à valoir sur le marché
„	**arent**	„ „ „ „ „ en cours
„	**arere**	avec **marge** de
„	**ares**	laisse(nt) trop peu de marge
„	**aretis**	notre (ma) **marque**
„	**aretur**	votre marque
„	**ari**	marque de fabrique
„	**arier**	„ supérieure
„	**at**	„ ordinaire
„	**atae**	marques supérieures
„	**atarum**	„ ordinaires
„	**atior**	premières marques
„	**atius**	autres marques
„	**atos**	télégraphiez marque
„	**atote**	télégraphiez marque et numéros
„	**atum**	nouvelle marque
„	**aturae**	même marque
„	**aturi**	marque usuelle
„	**aturos**	marquez comme de coutume
„	**averam**	en masse
„	**averit**	la **masse** active
„	**avero**	la masse passive
„	**avimus**	en **matière** de
„	**avisse**	la matière première
„	**avisti**	ce **matin**
„	**avit**	demain matin
„	**emus**	dans la matinée (de)
„	**ent**	il en est de **même** pour
„	**es**	même que précédemment
„	**etis**	pas même
„	**etur**	il est du **métier**
„	**or**	de son métier

inhi	**abam**	le **mètre** cube
„	**abamus**	le mètre carré
„	**abant**	faute de **mieux**
„	**abare**	pas moyen de faire mieux
„	**abaris**	il serait mieux
„	**abatur**	au **milieu** de
„	**abimur**	**mines** d'or
„	**abis**	télégraphiez **mode** (de)
„	**abitis**	de **moi**
„	**abo**	par „
„	**amini**	pour „
„	**amur**	sur „
„	**andam**	tout au **moins**
„	**andi**	beaucoup moins
„	**andior**	en moins
„	**andius**	à moins de
„	**andos**	„ „ que vous ne
„	**ans**	„ „ „ nous „
„	**antem**	moins que précédemment
„	**antis**	moins qu'à l'ordinaire
„	**antium**	par **mois**
„	**anto**	1 mois
„	**antur**	2 „
„	**aremus**	3 „
„	**arent**	4 „
„	**arere**	5 „
„	**ares**	6 „
„	**aretis**	9 „
„	**aretur**	**moitié** par moitié
„	**ari**	plus de la moitié
„	**arier**	en ce **moment**
„	**at**	pas en ce moment
„	**atae**	pour le moment
„	**atarum**	au moment de
„	**atior**	d'un moment à l'autre
„	**atius**	à tout moment
„	**atos**	au moment donné
„	**atote**	le moment actuel est propice (pour)
„	**atum**	le moment actuel n'est pas propice (pour)
„	**aturae**	télégraphiez lorsque le moment vous semblera propice
„	**aturi**	y compris le **montage**
„	**aturos**	montage non compris
„	**averam**	**mort** de
„	**averit**	„ „ fièvre
„	**avero**	„ „ choléra
„	**avimus**	on signale la mort de
„	**avisse**	par suite de la mort de
„	**avisti**	en simples **mots**
„	**avit**	télégraphiez en simples mots
„	**emus**	„ **motif** de
„	**ent**	il n'y a pas **moyen** de
„	**es**	s'il n'y a pas moyen (de)
„	**etis**	en **moyenne**
„	**etur**	en dessous de la moyenne
„	**or**	„ dessus „ „ „

intr	**abam**	a fait **naufrage**
„	**abamus**	**navigation** est ouverte
„	**abant**	„ „ fermée
„	**abare**	télégraphiez si navigation ouverte
„	**abaris**	à la réouverture de navigation
„	**abatur**	jusqu'à clôture de navigation
„	**abimur**	avant „ „ „
„	**abis**	**navire** en fer
„	**abitis**	„ „ bois
„	**abo**	si nécessaire
„	**amini**	pas nécessaire
„	**amur**	si pas nécessaire
„	**andam**	télégraphiez si nécessaire
„	**andi**	absolument nécessaire
„	**andior**	il est nécessaire
„	**andius**	il n'est pas nécessaire
„	**andos**	soignerons (soignerai) le nécessaire
„	**ans**	soignez le nécessaire
„	**antem**	soigné „ „
„	**antis**	par suite de la nécessité de
„	**antium**	**négativement**
„	**anto**	entamez des **négociations**
„	**antur**	sommes (suis) en négociations (pour)
„	**aremus**	négociations rompues
„	**arent**	„ reprises
„	**arere**	par suite de **neige**
„	**ares**	**net** sans escompte
„	**aretis**	il a fallu **refaire** à **neuf**
„	**aretur**	15 **noeuds**
„	**ari**	16 „
„	**arier**	17 „
„	**at**	20 „
„	**atae**	21 „
„	**atarum**	télégraphiez **nom** (de)
„	**atior**	„ „ et adresse (de)
„	**atius**	au nom de qui?
„	**atos**	à notre nom
„	**atote**	à votre nom
„	**atum**	à leur (son) nom
„	**aturae**	à mon nom personnel
„	**aturi**	nom inconnu
„	**aturos**	télégraphiez nom du bateau
„	**averam**	en grand **nombre**
„	**averit**	en petit nombre
„	**avero**	télégraphiez nombre (de) [(de)
„	**avimus**	„ „ approximatif
„	**avisse**	plus nombreux [nible
„	**avisti**	purement **nominal** faute de dispo-
„	**avit**	purement nominal faute d'ache-
„	**emus**	valeur nominale [teurs
„	**ent**	ou **non**
„	**es**	si non
„	**etis**	**Nord**-Est
„	**etur**	Nord-Ouest
„	**or**	par **nous**

iter	**abam**	de nous
„	**abamus**	à nous
„	**abant**	pour nous
„	**abare**	sur nous
„	**abaris**	de **nouveau**
„	**abatur**	à nouveau
„	**abimur**	envoyez-en un nouveau
„	**abis**	rien de nouveau
„	**abitis**	télégraphiez **nouvelles**
„	**abo**	sans nouvelles
„	**amini**	nouvelle dénudée de fondement
„	**amur**	d'après nouvelles dignes de foi
„	**andam**	d'après nouvelles reçues
„	**andi**	passagers **noyés**
„	**andior**	capitaine noyé
„	**andius**	dans la **nuit**
„	**andos**	la nuit passée
„	**ans**	toute la nuit
„	**antem**	à peu près **nul**
„	**antis**	nul et non avenu
„	**antium**	télégraphiez **numéro**
„	**anto**	pour satisfaire aux **obligations**
„	**antur**	satisfair(a)ont à ses (leurs) obliga-
„	**aremus**	obligations [tions
„	**arent**	sommes (suis) obligé(s) de
„	**arere**	il faut s'obliger à
„	**ares**	**obtenable** à
„	**aretis**	pas obtenable
„	**aretur**	si obtenable
„	**ari**	à l'**occasion** de
„	**arier**	à la prochaine occasion
„	**at**	il y a une occasion
„	**atae**	„ „ „ „ „ directe
„	**atarum**	„ „ „ „ „ favorable
„	**atior**	saisissez la première occasion (pour)
„	**atius**	il n'y a pas d'occasion
„	**atos**	„ „ „ „ d'autre occasion
„	**atote**	c'est un prix d'occasion
„	**atum**	**octroi** compris
„	**aturae**	„ non compris
„	**aturi**	combien pouvez-vous **offrir?**
„	**aturos**	nous v. faisons (je v. fais) offre ferme pour réponse télégraph. immédiate
„	**averam**	nous v. faisons (je v. fais) offre d'occasion pour réponse télégraph. immédiate
„	**averit**	on nous (m') offre
„	**avero**	offre sans engagement
„	**avimus**	nous vous offrons (je vous offre) jusqu' à révocation
„	**avisse**	nous vous offrons (je vous offre) sans engagement
„	**avisti**	tâchez de faire offre ferme
„	**avit**	„ „ proroger offre
„	**emus**	„ „ renouveler offre
„	**ent**	vous offrons (v. offre) sauf vente
„	**es**	sur offre ferme il sera probablement possible de faire mieux
„	**etis**	offre annulée
„	**etur**	„ acceptée
„	**or**	

marché- p. 36

jact	**abam**	offre renouvelée
„	**abamus**	„ refusée
„	**abant**	si impossible accepter offre, tâchez de faire contre-offre
„	**abare**	contre-offre pour réponse télégr. immédiate
„	**abaris**	renouvelons (renouvelle) offre
„	**abatur**	la concurrence offre
„	**abimur**	„ „ sollicite des offres
„	**abis**	télégraphiez prix à offrir
„	**abitis**	acceptons (accepte) offre et attendons (attends) confirmation
„	**abo**	acceptons (accepte) offre, sauf réponse contraire par télégr. immediat l'affaire sera considérée conclue
„	**amini**	offrons (offre) sur échantillon
„	**amur**	à **option**
„	**andam**	option double quantité
„	**andi**	déclarez option
„	**andior**	déclaré „
„	**andius**	option d'annuler
„	**andos**	„ à déclarer
„	**ans**	votre option
„	**antem**	notre (mon) option
„	**antis**	option de l'acheteur
„	**antium**	„ du vendeur
„	**anto**	suivant option
„	**antur**	„ „ annulons
„	**aremus**	en **or**
„	**arent**	d'or
„	**arere**	or en barres
„	**ares**	onces d'or
„	**aretis**	par suite d'**orages**
„	**aretur**	comme d'**ordinaire**
„	**ari**	de premier **ordre**
„	**arier**	d'ordre secondaire
„	**at**	en ordre
„	**atae**	pas en ordre
„	**atarum**	télégraphiez si en ordre
„	**atior**	à ordre
„	**atius**	à l'ordre de
„	**atos**	à votre ordre
„	**atote**	à notre (mon) ordre
„	**atum**	ou ordre
„	**aturae**	attendez ordres ultérieurs
„	**aturi**	déclarez ordre
„	**aturos**	sur l'ordre de
„	**averam**	sauf ordre contraire de votre [part
„	**averit**	sauf ordre contraire de notre (ma) [part
„	**avero**	télégraphiez ordres
„	**avimus**	tous les ordres
„	**avisse**	joignez à l'ordre en note
„	**avisti**	suivant ordre
„	**avit**	annulez ordre (No.)
„	**emus**	annulez tous les ordres en note
„	**ent**	annulé ordre (No.)
„	**es**	annulé tous les ordres en note
„	**etis**	
„	**etur**	
„	**or**	

jug	**abam**	par ordre et pour compte (de)
„	**abamus**	pas d'ordre en note
„	**abant**	tous les ordres sont exécutés
„	**abare**	l'ordre s'annule, n'ayant pas été exécuté à temps
„	**abaris**	solde de l'ordre
„	**abatur**	avez-vous reçu ordre?
„	**abimur**	nous avons (j'ai) reçu ordre
„	**abis**	ordre pas reçu
„	**abitis**	télégraphiez ordre ferme
„	**abo**	sur ordre ferme
„	**amini**	ordre plus en vigueur
„	**amur**	ordre d'essai
„	**andam**	débordé(s) d'ordres
„	**andi**	nous sommes (je suis) débordé/s d'ordres
„	**andior**	les ordres doivent être accompagnés de remises
„	**andius**	suspendez exécution ordre
„	**andos**	vu la situation critique du commettant suspendez exécution ordre
„	**ans**	télégraphiez **origine**
„	**antem**	télégraphiez **où**
„	**antis**	par où
„	**antium**	d'où
„	**anto**	où en êtes-vous de?
„	**antur**	si **oui**
„	**aremus**	par suite d'**ouragan**
„	**arent**	à l'**ouverture**
„	**arere**	dès l'ouverture
„	**ares**	avant ouverture
„	**aretis**	au **pair**
„	**aretur**	sous **palan**
„	**ari**	en **papier** monnaie
„	**arier**	à la **parité**
„	**at**	au-dessus de parité
„	**atae**	„ -dessous de parité
„	**atarum**	à **part**
„	**atior**	de part et d'autre
„	**atius**	à **partager** par moitié
„	**atos**	partageons la différence
„	**atote**	bâteau **partira**
„	**atum**	„ „ ce matin
„	**aturae**	„ „ ce soir
„	**aturi**	„ „ demain
„	**aturos**	„ parti
„	**averam**	„ „ ce matin
„	**averit**	„ „ ce soir
„	**avero**	quand partira (partiront)
„	**avimus**	**particulièrement**
„	**avisse**	en **partie**
„	**avisti**	en entier ou en partie
„	**avit**	
„	**emus**	
„	**ent**	
„	**es**	
„	**etis**	
„	**etur**	
„	**or**	

laet	abam	au moins en partie
„	abamus	en parties de
„	abant	la majeure partie
„	abare	**pas** de
„	abaris	si pas de
„	abatur	en **passage**
„	abimur	passage 1ère classe
„	abis	passage 2ième classe
„	abitis	passage 3ième classe
„	abo	réservez passage 1ère classe
„	amini	„ „ 2ième „
„	amur	réservé passage 1ère classe
„	andam	„ „ 2ième „
„	andi	comme par le **passé**
„	andior	sous **pavillon**
„	andius	**payement** partiel
„	andos	avis de payement
„	ans	„ „ „ pas encore reçu
„	antem	à valoir sur payement
„	antis	„ „ „ payements
„	antium	solde à payer
„	anto	payement comptant
„	antur	„ comptant lors de commande
„	aremus	„ „ contre expédition
„	arent	„ 2/3 „ lors de commande
„	arere	payement 2/3 comptant lors de commande et solde contre expédition
„	ares	payement 3/4 comptant lors de commande
„	aretis	payement 3/4 comptant lors de commande et solde contre expédition
„	aretur	paye(nt) régulièrement
„	ari	„ lentement
„	arier	payement à l'avance
„	at	payez „ „
„	atac	payé „ „
„	atarum	payement chèque
„	atior	payez „
„	atius	payé „
„	atos	payement à compte
„	atote	payez „ „
„	atum	payé „ „
„	aturae	payement en entier
„	aturi	payé „ „
„	aturos	payement moyennant
„	averam	„ à la banque
„	averit	payez „ „ „
„	avero	payé „ „ „
„	avimus	payement immédiat
„	avisse	sauf payement immédiat
„	avisti	payez immédiatement
„	avit	payement de traite
„	emus	„ du solde
„	ent	payez solde
„	es	traite payée
„	etis	solde payé
„	etur	payement sous protêt
„	or	payez sous protêt

lani	abam	payé sous protêt
„	abamus	payement contre rembourse-
„	abant	facture payée [ment
„	abare	payable à la banque
„	abaris	„ en acomptes
„	abatur	„ contre document(s)
„	abimur	faute payement
„	abis	en cas de non-payement
„	abitis	télégraphiez mode de paye-
„	abo	à qui faut-il payer? [ment
„	amini	effet impayé
„	amur	„ „ faute de
„	andam	assez pour payer
„	andi	pas assez pour payer
„	andior	suspendu payements
„	andius	repris payements
„	andos	présentez au payement
„	ans	présenté „ „ [ment
„	antem	garantit (garantissent) paye-
„	antis	garantis (garantissons) paye-
„	antium	à **peine** [ment
„	anto	il y a à peine
„	antur	**perçu** en trop
„	aremus	perçu en moins
„	arent	s'est **perdu** corps et biens
„	arere	tout l'équipage péri
„	ares	tous à bord péri
„	aretis	**permis** d'embarquement
„	aretur	contre permis d'embarque-
„	ari	obtenez **permission** [ment
„	arier	obtenu permission nécessaire
„	at	en **perspective**
„	atac	sur la pérspective de
„	atarum	les perspectives sont excellentes
„	atior	„ „ „ bonnes
„	atius	„ „ „ médiocres
„	atos	„ „ „ mauvaises
„	atote	en **perte**
„	atum	par suite de pertes
„	aturae	„ „ „ „ sensibles
„	aturi	éprouvé pertes
„	aturos	„ „ sensibles
„	averam	il y a une perte de
„	averit	en cas de perte
„	avero	sans perte
„	avimus	avec perte (de)
„	avisse	les pertes sont
„	avisti	„ „ „ sensibles
„	avit	„ „ „ médiocres
„	emus	„ „ „ nulles
„	ent	**peu** à peu
„	es	peu de
„	etis	trop peu (de)
„	etur	c' est trop peu
„	or	

larv	**abam**	il y a peu de
„	**abamus**	sous **peu**
„	**abant**	vu le peu de
„	**abare**	sur **place**
„	**abaris**	sur votre place
„	**abatur**	sur notre place
„	**abimur**	sur leur place
„	**abis**	pas de place
„	**abitis**	peu de place
„	**abo**	on se **plaint** de
„	**amini**	sous **pli** cacheté et scellé
„	**amur**	les **pluies** tombent
„	**andam**	par suite de pluies
„	**andi**	par suite du manque de pluies
„	**andior**	pour la **plupart**
„	**andius**	la plupart des
„	**andos**	tout au **plus**
„	**ans**	de plus
„	**antem**	plus ou moins
„	**antis**	5 % de plus ou moins
„	**antium**	10 „ „ „ „ „
„	**anto**	15 „ „ „ „ „
„	**antur**	plus de
„	**aremus**	si plus (de)
„	**arent**	plus que précédemment
„	**arere**	plus que stipulé
„	**ares**	plus qu'anticipé
„	**aretis**	au plus tôt
„	**aretur**	au **poids**
„	**ari**	poids délivré
„	**arier**	sans garantie de poids
„	**at**	bulletin de poids
„	**atae**	suivant bulletin de poids
„	**atarum**	envoyez „ „ „
„	**atior**	suivant poids
„	**atius**	poids net
„	**atos**	poids brut
„	**atote**	poids approximatif
„	**atum**	poids minimum
„	**aturae**	poids maximum
„	**aturi**	poids constaté
„	**aturos**	poids garanti
„	**averam**	poids moyen
„	**averit**	poids total
„	**avero**	télégraphiez poids brut
„	**avimus**	„ poids net
„	**avisse**	„ „ approximatif
„	**avisti**	„ „ constaté
„	**avit**	„ „ moyen
„	**emus**	„ „ total
„	**ent**	poids brut et net [colis
„	**es**	„ „ „ „ de chaque
„	**etis**	indiquez poids brut et net de chaque article
„	**etur**	indiquez poids brut et net de chaque colis
„	**or**	il y a différence de poids en plus

lass	**abam**	il y a différence de poids en moins
„	**abamus**	il y a un déficit sensible de poids
„	**abant**	par suite de déficit de poids
„	**abare**	par suite de déficit sensible de poids
„	**abaris**	bonification pour déficit de poids
„	**abatur**	on exige bonification pour déficit de poids
„	**abimur**	marchandise trop lourde
„	**abis**	„ „ légère
„	**abitis**	nous sommes (je suis) sur le
„	**abo**	au point de vue de [**point** de
„	**amini**	à tous les points de vue
„	**amur**	il y a des rumeurs **politiques**
„	**andam**	par suite de complications
„	**andi**	en **port** dû [politiques
„	**andior**	en port payé
„	**andius**	franco de port
„	**andos**	port d'embarquement
„	**ans**	port de débarquement
„	**antem**	port d'escale
„	**antis**	port de salut
„	**antium**	port d'ordres
„	**anto**	au **porteur**
„	**antur**	même **position**
„	**aremus**	vu la **possibilité** de
„	**arent**	encore possible
„	**arere**	si possible
„	**ares**	si encore possible
„	**aretis**	télégraphiez si possible
„	**aretur**	télégraphiez si encore possible
„	**ari**	plus possible
„	**arier**	si possible, télégraphiez
„	**at**	si encore possible, télégraphiez
„	**atae**	„ plus „ „
„	**atarum**	possible(s) mais pas probable(s)
„	**atior**	il est encore possible (de)
„	**atius**	il n'est plus „ „
„	**atos**	aussitôt que possible
„	**atote**	par la **poste**
„	**atum**	poste restante
„	**aturae**	envoyez par la poste
„	**aturi**	envoyé „ „ „
„	**aturos**	voilà **pourquoi**
„	**averam**	télégraphiez pourquoi
„	**averit**	**poursuivez** en payement
„	**avero**	poursuivr/ons(ai) en payement
„	**avimus**	il y a eu des poursuites
„	**avisse**	poursuites ont été fréquentes
„	**avisti**	il n'y a pas eu de poursuites
„	**avit**	bien **pourvu**
„	**emus**	encore „
„	**ent**	mal „
„	**es**	pourvu que
„	**etis**	pourvu que vous
„	**etur**	„ „ nous (je)
„	**or**	

laud	**abam**	pourvu qu'on
„	**abamus**	fondé de **pouvoirs**
„	**abant**	plein pouvoir
„	**abare**	pouvoir spécial
„	**abaris**	envoyez pouvoir
„	**abatur**	„ plein pouvoir
„	**abimur**	„ pouvoir spécial
„	**abis**	„ formule de pouvoir
„	**abitis**	envoyé pouvoir
„	**abo**	pouvoir au nom de
„	**amini**	envoyez pouvoir en notre nom
„	**amur**	„ „ „ mon „
„	**andam**	pouvoir notarié
„	**andi**	avec plein pouvoir
„	**andior**	ne peut (ne peuvent)
„	**andius**	nous pouvons (je peux)
„	**andos**	pouvez-vous?
„	**ans**	libre **pratique**
„	**antem**	au **préalable**
„	**antis**	toutes les **précautions** ont été prises
„	**antium**	prenez toutes vos précautions
„	**anto**	**précédemment**
„	**antur**	que précédemment
„	**aremus**	plus que précédemment
„	**arent**	moins „ „
„	**arere**	même „ „
„	**ares**	tout ce qui précède
„	**aretis**	s'applique à tout ce qui précède
„	**aretur**	il est **préférable**
„	**ari**	si préférable
„	**arier**	télégraphiez si vous jugez préférable
„	**at**	il semble préférable
„	**atae**	par préférence
„	**atarum**	à égalité de prix aurez la préférence
„	**atior**	préférez-vous?
„	**atius**	on préfère
„	**atos**	que préférez-vous?
„	**atote**	si vous préférez
„	**atum**	au **préjudice** de
„	**aturae**	à notre préjudice
„	**aturi**	à mon „
„	**aturos**	à votre „
„	**averam**	à leur (son) „
„	**averit**	télégraphiez **prénom**
„	**avero**	**prépond**érance
„	**avimus**	il y a prépondérance (de)
„	**avisse**	à peu **près**
„	**avisti**	en **présence** de
„	**avit**	en notre présence
„	**emus**	„ ma „
„	**ent**	„ votre „
„	**es**	jusqu'à **présent**
„	**etis**	comme jusqu'à présent
„	**etur**	présentation
„	**or**	à la présentation

lent	**abam**	très **pressé**(s)
„	**abamus**	pas „
„	**abant**	si „
„	**abare**	tout est **prêt**
„	**abaris**	pas encore „
„	**abatur**	quand sera „
„	**abimur**	sous **prétexte** (de)
„	**abis**	inférieur aux **prévisions**
„	**abitis**	supérieur „ „
„	**abo**	il est impossible de **prévoir**
„	**amini**	on prévoit
„	**amur**	fonds de **prévo**yance
„	**andam**	télégraphiez **prime**
„	**andi**	**principal** et frais
„	**andior**	en **principe**
„	**andius**	télégraphiez avis sur **prix** futurs
„	**andos**	prix ont atteint le plus haut niveau
„	**ans**	prix ont atteint le plus bas niveau
„	**antem**	télégraphiez prix qualités
„	**antis**	„ „ pour toutes les
„	**antium**	télégraphiez prix pour tous les articles
„	**anto**	télégraphiez tout dernier prix
„	**antur**	télégraphiez tout dernier prix et plus bref délai livraison
„	**aremus**	prix nominaux faute d'acheteurs
„	**arent**	„ „ „ de vendeurs
„	**arere**	prix susceptibles de fluctuations
„	**ares**	prix sont stationnaires
„	**aretis**	mêmes prix pour
„	**aretur**	ce prix est très avantageux
„	**ari**	prix actuels
„	**arier**	„ de la concurrence
„	**at**	„ de convention
„	**atae**	„ de contrat
„	**atarum**	„ cotés
„	**atior**	„ coûtant
„	**atius**	„ demandés
„	**atos**	„ de facture
„	**atote**	„ de faveur
„	**atum**	„ offerts
„	**aturae**	„ payés
„	**aturi**	„ le plus réduit possible
„	**aturos**	„ les plus réduits possible
„	**averam**	à ce prix
„	**averit**	à ces prix
„	**avero**	au prix actuel
„	**avimus**	„ „ de la concurrence
„	**avisse**	„ „ „ contrat
„	**avisti**	„ „ coûtant
„	**avit**	au tout dernier prix
„	**emus**	c'est un prix de faveur
„	**ent**	au prix le plus favorable
„	**es**	„ „ offert
„	**etis**	aux „ actuels
„	**etur**	„ „ de la concurrence
„	**or**	„ „ de convention

lig	abam		aux prix de contrat
„	abamus		„ „ coûtant
„	abant		„ „ demandés
„	abare		„ „ de faveur
„	abaris		„ „ les plus favorables
„	abatur		„ „ précédents
„	abimur		„ „ de revient
„	abis		à prix fixes
„	abitis		„ „ réduits
„	abo		à prix plus élevé
„	amini		„ „ „ élevés
„	amur		sur la base de ce prix
„	andam		„ „ „ „ ces „
„	andi		les prix s'entendent
„	andior		télégraphiez comment prix s'entend(ent)
„	andius		achats considérables
„	andos	prix	découvert considérable
„	ans	plus	faibles arrivages
„	antem	élevés	forte demande
„	antis		nombreuses demandes
„	antium	par	flottant insignifiant
„	anto	suite	frêts plus élevés
„	antur	d/e	ordres d'achats
„	aremus		rachats
„	arent		diminution de stocks
„	arere		stocks insignifiants
„	ares		demande insignifiante
„	aretis	prix	diminution de demande
„	aretur	plus	expéditions considérables
„	ari	bas	frêts plus bas
„	arier		nombreuses offres
„	at	par	ordres de vente
„	atae	suite	forte production
„	atarum	d/e	réalisations
„	atior		récolte plus favorable
„	atius		stocks considérables
„	atos		prix-courant
„	atote		prix-courants
„	atum		au prix-courant
„	aturae		aux prix-courants
„	aturi		envoyez prix-courant
„	aturos		„ prix-courants
„	averam		nouveau prix-courant
„	averit		ancien „ „
„	avero		nouveaux prix-courants
„	avimus		anciens „ „
„	avisse		nouveau prix-courant en vigueur
„	avisti		nouveaux prix-courants en vigueur
„	avit		suivant nouveau prix-courant
„	emus		„ nouveaux prix-courants
„	ent		„ ancien prix-courant
„	es		„ anciens prix-courants
„	etis		**probablement**
„	etur		vu la probabilité (de)
„	or		selon toutes les probabilités

line	abam	il y a probabilité de
„	abamus	frais de **procès**
„	abant	télégraphiez estimation frais de procès
„	abare	procès verbal
„	abaris	dressez procès verbal
„	abatur	par **procuration**
„	abimur	**production** totale
„	abis	surplus de production
„	abitis	surproduction
„	abo	par suite de surproduction
„	amini	toute la production est engagée
„	amur	**produit** net
„	andam	télégraphiez produit net
„	andi	à valoir sur produit net
„	andior	**prolongation**
„	andius	en **proportion** égale
„	andos	hors de proportion
„	ans	proportionné à
„	antem	**proposition**
„	antis	si impossible d'accepter proposition, faites contre-proposition
„	antium	tâchez d'obtenir **prorogation**
„	anto	à la protestation en cas de refus
„	antur	frais de **protêt**
„	aremus	sans protêt
„	arent	traite protestée faute d'acceptation
„	arere	„ „ „ de payement
„	ares	par suite de protêt
„	aretis	sous protêt
„	aretur	évitez protêt
„	ari	en cas de protêt
„	arier	acceptons sous protêt
„	at	accepté „ „
„	atae	télégraphiez **provenance**
„	atarum	garanti **pur**
„	atior	télégraphiez **qualité**
„	atius	sous le rapport de la qualité
„	atos	même qualité
„	atote	„ „ et quantité
„	atum	„ „ quantité et prix
„	aturae	mêmes qualités
„	aturi	„ „ et quantités
„	aturos	si qualité convient
„	averam	qualité laisse à désirer
„	averit	vu l'infériorité de qualité
„	avero	bonification pour infériorité de qualité
„	avimus	on exige bonification pour infériorité de qualité
„	avisse	envoi refusé à cause d'infériorité de qualité
„	avisti	retournons envoi qualité étant inférieure
„	avit	on se plaint de la qualité
„	emus	qualité égale à
„	ent	„ „ à l'échantillon
„	es	„ „ aux échantillons
„	etis	„ „ à notre échantillon
„	etur	„ „ „ votre „
„	or	„ à peu près égale à

liqu	**abam**	qualité à peu près égale à échantillon(s)
„	**abamus**	toute première qualité
„	**abant**	première qualité
„	**abare**	2ième qualité
„	**abaris**	qualité de choix
„	**abatur**	„ supérieure
„	**abimur**	„ inférieure
„	**abis**	„ irréprochable
„	**abitis**	„ garantie
„	**abo**	meilleure qualité possible
„	**amini**	bonne qualité
„	**amur**	qualité courante
„	**andam**	„ moyenne
„	**andi**	„ ordinaire
„	**andior**	„ habituelle
„	**andius**	„ médiocre
„	**andos**	„ intermédiaire
„	**ans**	„ convenable
„	**antem**	„ loyale et marchande
„	**antis**	autre qualité
„	**antium**	autres qualités
„	**anto**	toutes les qualités
„	**antur**	qualité inutilisable
„	**aremus**	qualité et conditionnement
„	**arent**	même qualité que la fois passée
„	**arere**	télégraphiez **quand**
„	**ares**	jusqu'à quand?
„	**aretis**	quand même
„	**aretur**	télégraphiez quand approximativement
„	**ari**	télégraphiez **quantité**
„	**arier**	quantité totale
„	**at**	„ moyenne
„	**atae**	„ maximum
„	**atarum**	„ minimum
„	**atior**	„ achetée
„	**atius**	„ livrée
„	**atos**	„ vendue
„	**atote**	télégraphiez quantité
„	**atum**	„ „ approximative
„	**aturae**	„ „ totale
„	**aturi**	„ „ maximum
„	**aturos**	„ „ minimum
„	**averam**	„ „ achetée
„	**averit**	„ „ embarquée
„	**avero**	„ „ vendue
„	**avimus**	à valoir sur la quantité
„	**avisse**	petite quantité
„	**avisti**	grande quantité
„	**avit**	quantité plus petite
„	**emus**	„ „ grande
„	**ent**	quantités importantes
„	**es**	augmentez la quantité à
„	**etis**	tâchez d'augmenter la quantité à
„	**etur**	quantités plus petites proportionnellement plus cher
„	**or**	quantités plus grandes proportionnellement meilleur marché

lucr	**abam**	quantité est trop petite
„	**abamus**	quantité est trop grande
„	**abant**	„ complémentaire
„	**abare**	**quarantaine**
„	**abaris**	quarantaine a été imposée
„	**abatur**	en quarantaine
„	**abimur**	par suite de quarantaine
„	**abis**	quarantaine est levée
„	**abitis**	détenu par quarantaine
„	**abo**	envoyez **quittance**
„	**amini**	contre quittance
„	**amur**	5 % de **rabais**
„	**andam**	10 % „ „
„	**andi**	15 % „ „
„	**andior**	20 % „ „
„	**andius**	25 % „ „
„	**andos**	30 % „ „
„	**ans**	33¹/₃ % de „
„	**antem**	40 % de „
„	**antis**	45 % „ „
„	**antium**	50 % „ „
„	**anto**	télégraphiez taux de rabais
„	**antur**	point de rabais
„	**aremus**	par suite de **rachats**
„	**arent**	sous la **raison** de
„	**arere**	à raison de
„	**ares**	de 1er **rang**
„	**aretis**	envoyez **rapport**
„	**aretur**	d'après des rapports (de)
„	**ari**	il y a des rapports (de)
„	**arier**	envoyez rapport détaillé
„	**at**	rapport détaillé par lettre de ce jour
„	**atae**	rapport plus détaillé suivra
„	**atarum**	**rapproché** et dû
„	**atior**	très **rare**
„	**atius**	rare et ferme
„	**atos**	par suite de la rareté de
„	**atote**	**ravitaillement**
„	**atum**	en **réalisation** de bénéfice
„	**aturae**	il y a eu des réalisations
„	**aturi**	par suite de réalisations
„	**aturos**	montant réalisé
„	**averam**	en **réalité**
„	**averit**	contre **récépissé**
„	**avero**	à **réception**
„	**avimus**	recettes totales
„	**avisse**	**réclamation**
„	**avisti**	réclamations
„	**avit**	par suite de réclamation
„	**emus**	il y a de nombreuses réclamations
„	**ent**	en cas de réclamation
„	**es**	réclamation pas justifiée
„	**etis**	„ est „
„	**etur**	**récolte** dernière
„	**or**	récolte actuelle

prix— p. 42

lux	abam	dernière et nouvelle récoltes
„	abamus	récolte future
„	abant	télégraphiez conditionnement de récolte
„	abare	télégraphiez estimation de la récolte
„	abaris	surplus de récolte disponible pour l'exportation
„	abatur	les semailles sont tardives
„	abimur	superficie ensemencée
„	abis	récolte commence à battre son plein
„	abitis	récolte est rentrée
„	abo	récolte pendante
„	amini	la cueillette est commencée
„	amur	**recommandation**
„	andam	**recommencement**
„	andi	pas **reçu** ce que vous aviez envoyé
„	andior	sans **réduction**
„	andius	à une réduction
„	andos	par suite de réduction
„	ans	**réduit** à
„	antem	en cas de **refus**
„	antis	par suite du refus de
„	antium	droits de **régie**
„	anto	dans toutes les **régions**
„	antur	en **règle**
„	aremus	télégraphiez si en règle
„	arent	si pas en règle, télégraphiez
„	arere	si en règle
„	ares	pas en règle
„	aretis	sauf **règlement** immédiat
„	aretur	règlement au moyen de
„	ari	jour de règlement
„	arier	règlement de compte
„	at	télégraphiez mode de règlement
„	atae	au prochain règlement
„	atarum	en **relation** (avec)
„	atior	évitez toute relation (avec)
„	atius	mettez-vous en relation avec
„	atos	**remboursable**
„	atote	remboursement
„	atum	contre remboursement
„	aturae	délivrez contre remboursement
„	aturi	expédiez „ „
„	aturos	fait suivre en remboursement
„	averam	à **remettre** jusqu'à
„	averit	remis jusqu'à
„	avero	remettez autant que possible
„	avimus	remettez tous les fonds disponibles
„	avisse	contre remise
„	avisti	à valoir sur remise
„	avit	remettez par transfert télégraphique
„	emus	remis „ „ „
„	ent	**remplacement**
„	es	„ gratuit
„	etis	en replacement (de)
„	etur	il faut remplacer gratuitement
„	or	remplacerons (remplacerai) gratuitement

man	abam	télégraphiez **rendement**
„	abamus	**renseignements**
„	abant	télégraphiez renseignements (sur)
„	abare	d'après renseignements reçus
„	abaris	renseignements reçus sont favorables
„	abatur	renseignements reçus sont très favorables
„	abimur	renseignements reçus sont défavorables
„	abis	renseignements reçus sont très défavorables
„	abitis	renseignements reçus sont contradictoires
„	abo	impossible obtenir renseignements
„	amini	avez-vous reçu renseignements?
„	amur	pas reçu renseignements
„	andam	envoyez renseignements par lettre
„	andi	renseignements complets par lettre de ce jour
„	andior	ces renseignements sont donnés sous toute réserve
„	andius	à **répartir** également
„	andos	**répondez** immédiatement par télégramme
„	ans	sauf réponse immédiate
„	antem	sauf réponse contraire par télégramme immédiat
„	antis	pas reçu réponse
„	antium	taux de **report**
„	anto	télégraphiez taux de report
„	antur	**représentant**
„	aremus	représentants
„	arent	représentation
„	arere	en **réserve**
„	ares	porté à la réserve
„	aretis	fonds de réserve
„	aretur	sans réserve
„	ari	sous toutes réserves
„	arier	sous réserve
„	at	sous la réserve usitée
„	atae	très réservé
„	atarum	**responsable**
„	atior	pas responsable
„	atius	décline(nt) toute responsabilité
„	atos	sans responsabilité
„	atote	„ „ de notre (ma) part
„	atum	„ „ „ votre part
„	aturae	il **reste** (à)
„	aturi	télégraphiez ce qui reste (à)
„	aturos	il ne reste plus que
„	averam	**restreignez** autant que possible
„	averit	**résultant** de
„	avero	télégraphiez **résultat**
„	avimus	résultat n'est pas encore connu
„	avisse	„ est encore incertain
„	avisti	„ sera probablement favorable
„	avit	résultat sera probablement défavorable
„	emus	
„	ent	
„	es	
„	etis	
„	etur	
„	or	

mersabam	aussitôt connu, le résultat vous sera télégraphié
„ abamus	à quel résultat vous attendez-vous?
„ abant	sans résultat
„ abare	résultat est favorable
„ abaris	„ „ passable
„ abatur	„ „ défavorable
„ abimur	„ „ excellent
„ abis	par suite de résultat favorable
„ abitis	„ „ „ „ défavorable
„ abo	résultat probable
„ amini	résultat définitif
„ amur	télégraphiez résultat probable
„ andam	„ „ définitif
„ andi	sans **retard**
„ andior	beaucoup en retard
„ andius	retardé par suite de
„ andos	par suite de retard
„ ans	il y aura du retard
„ antem	en cas de retard
„ antis	évitez tout retard
„ antium	bonification pour retard de livraison
„ anto	on exige pour retard de livraison bonification (de)
„ antur	envoi mis à disposition à cause du retard de livraison
„ aremus	transaction annulée par suite de retard
„ arent	**retournez** la marchandise
„ arere	marchandise retournée
„ ares	marchandise retournée à cause de retard de livraison
„ aretis	marchandise retournée à cause de qualité inférieure
„ aretur	retournez ce qui vous a été envoyé
„ ari	retourné ce qui nous (m') a été envoyé
„ arier	parfaitement **réussi**
„ at	pas réussi
„ atae	par suite de **reventes**
„ atarum	**révocable** 3 mois à l'avance
„ atior	„ 6 „ „ „
„ atius	jusqu'à révocation
„ atos	**rien** du tout
„ atote	à la **rigueur**
„ atum	vu les **risques**
„ aturae	il semble risqué
„ aturi	il ne semble pas risqué
„ aturos	en **route** (pour)
„ averam	d'après **rumeurs** répandues
„ averit	en **sacs**
„ avero	sacs compris
„ avimus	sacs perdus
„ avisse	sacs de location
„ avisti	reprise des sacs à
„ avit	la **saison** a commencé
„ emus	„ „ commencera
„ ent	„ „ est en retard
„ es	„ „ tire à sa fin
„ etis	„ „ sera terminée
„ etur	„ „ est „
„ or	**salaire** mensuel

miniabam	salaire annuel
„ abamus	salaire augmenté à
„ abant	pour raison de **santé**
„ abare	télégraphiez état de santé
„ abaris	„ „ „ votre santé
„ abatur	état de santé est excellent
„ abimur	„ „ „ „ bon
„ abis	„ „ „ „ mauvais
„ abitis	„ „ „ „ critique
„ abo	„ „ „ „ meilleur
„ amini	„ „ „ „ plus mauvais
„ amur	difficile à **satisfaire**
„ andam	tous **sauvés**
„ andi	par suite de **sécheresse**
„ andior	par **semaine**
„ andius	une fois par semaine
„ andos	première semaine (de)
„ ans	**semblable** à
„ antem	à ce qu'il **semble**
„ antis	il nous (me) semble que
„ antium	il ne semble pas que
„ anto	prochain **semestre**
„ antur	semestre en cours
„ aremus	pendant le semestre en cours
„ arent	dans le **sens** indiqué
„ arere	**sentence** prononcée contre
„ ares	appelez de la sentence
„ aretis	en **service** de groupage
„ aretur	expédiez en service de groupage
„ ari	expédié „ „ „ „
„ arier	**signé** par
„ at	envoyez compte **simulé**
„ atae	suivant „ „
„ atarum	en cas de **sinistre**
„ atior	renseignez-nous (moi) sur la **situation**
„ atius	profitez de la situation
„ atos	la situation est favorable
„ atote	„ „ „ passable
„ atum	„ „ „ défavorable
„ aturae	„ „ „ critique
„ aturi	„ „ „ incertaine
„ aturos	„ „ „ sans changement
„ averam	„ „ s'est améliorée
„ averit	„ „ a empiré
„ avero	ce **soir**
„ avimus	demain soir
„ avisse	hier soir
„ avisti	**solde** en notre (ma) faveur
„ avit	solde en votre faveur
„ emus	télégraphiez solde
„ ent	„ „ en notre (ma) faveur
„ es	à valoir sur le solde
„ etis	remettez solde en notre (ma) faveur
„ etur	remis solde en votre faveur
„ or	sur la **sollicitation** de

mult	abam	aucune **solution** définitive n'est intervenue
„	abamus	**solvabilité**
„	abant	télégraphiez si et à concurrence de quel montant solvable
„	abare	toutes les **sortes**
„	abaris	autres sortes
„	abatur	quelle sorte?
„	abimur	quelles sortes?
„	abis	de sorte que
„	abitis	en **souffrance**
„	abo	reste en souffrance
„	amini	par voie de **soumission**
„	amur	gouvernement sollicite la soumission de
„	andam	prix de soumission
„	andi	**souscription**
„	andior	„ au pair
„	andius	suivant **spécification**
„	andos	envoyez „
„	ans	envoyé „
„	antem	**spéculateur(s)**
„	antis	spéculation
„	antium	pour spéculation
„	anto	par suite de spéculations
„	antur	il y a une **stagnation** (de)
„	aremus	franco **station**
„	arent	„ „ de cette place
„	arere	„ „ „ votre „
„	ares	stationnaire(s)
„	aretis	**statu**-quo
„	aretur	laissez les choses in statu-quo
„	ari	envoyez le **status**
„	arier	envoyé „ „
„	at	à **substituer** (à)
„	atae	en substitution de
„	atarum	avec **succès**
„	atior	sans succès
„	atius	en cas de succès
„	atos	**Sud**-Est
„	atote	Sud-Ouest
„	atum	**suffisamment**
„	aturae	par **suite** de
„	aturi	probablement par suite de
„	aturos	dans la suite
„	averam	tout ce qui **suit**
„	averit	reste suivra aussitôt possible
„	avero	au **sujet** de
„	avimus	à ce sujet
„	avisse	on **suppose** que
„	avisti	d'après ce qu'on suppose
„	avit	**surabondance** (de)
„	emus	par suite de surabondance (de)
„	ent	le **surlendemain**
„	es	en **sus**
„	etis	**suspension** de payements
„	etur	„ de payements est imminente
„	or	suspendu payements

narr	abam	en cas de suspension de payements
„	abamus	par suite de suspension
„	abant	par suite de suspension de payements
„	abare	d'après un bruit qui court, suspendu payements
„	abaris	d'après un bruit qui court, suspension de payements est imminente
„	abatur	suspendez jusqu'à nouvel ordre
„	abimur	suspendu momentanément
„	abis	totalement suspendu
„	abitis	en **suspens**
„	abo	restera (resteront) en suspens
„	amini	laissez en suspens
„	amur	trop **tard**
„	andam	au plus tard
„	andi	arrivé trop tard
„	andior	expédié trop tard
„	andius	plus tard
„	andos	pas plus tard (que)
„	ans	si trop tard
„	antem	si trop tard, télégraphiez
„	antis	ne **tardez** pas à
„	antium	nouveau **tarif**
„	anto	ancien tarif
„	antur	d'après nouveau tarif
„	aremus	d'après tarif actuel
„	arent	tarif en vigueur
„	arere	tarifs seront modifiés
„	ares	d'après nouveaux tarifs
„	aretis	d'après anciens tarifs
„	aretur	d'après tarifs en vigueur
„	ari	nouveau tarif en vigueur
„	arier	nouveaux tarifs en vigueur
„	at	d'après nouveau **tarif** en vigueur
„	atae	„ nouveaux tarifs „ „
„	atarum	télégraphiez **taux** pour
„	atior	au taux le plus favorable
„	atius	**tel** que
„	atos	tel quel
„	atote	**télégraphiez**
„	atum	„ si
„	aturae	„ aussitôt possible
„	aturi	„ immédiatement
„	aturos	„ réponse immédiate
„	averam	„ à
„	averit	par télégramme urgent
„	avero	suivant télégramme privé
„	avimus	„ „ officiel
„	avisse	on télégraphie de
„	avisti	votre télégramme est arrivé trop tard
„	avit	télégraphiez en simples mots
„	emus	télégraphiez plus clairement
„	ent	par **téléphone**
„	es	par suite d'une **tempête**
„	etis	en son **temps**
„	etur	
„	or	

nav	abam	en temps utile
„	abamus	en même temps
„	abant	combien de temps faut-il ?
„	abare	dans quelque temps
„	abaris	en ces derniers temps
„	abatur	d'ici à très peu de temps
„	abimur	vu la perte de temps
„	abis	pour un temps indéterminé
„	abitis	pour un temps à déterminer
„	abo	il est encore temps
„	amini	pendant qu'il en est encore temps
„	amur	est-il encore temps?
„	andam	s'il n'est plus temps
„	andi	„ „ „ „ télégraphiez
„	andior	de temps à autre
„	andius	à temps
„	andos	si encore à temps
„	ans	vu le peu de temps
„	antem	si le temps le permet
„	antis	**tendant** à
„	antium	fermement **tenu**
„	anto	faiblement tenu
„	antur	nous tenons à ce que
„	aremus	nous tenons à ne pas
„	arent	si vous tenez à
„	arere	à **terme**
„	ares	à court terme
„	aretis	à long terme
„	aretur	termes éloignés
„	ari	terme est trop court
„	arier	aussitôt **terminé**
„	at	„ „ télégraphiez
„	atae	pas encore terminé
„	atarum	télégraphiez quand se terminera
„	atior	muni du **timbre**
„	atius	papier timbré
„	atos	à **titre** de
„	atote	**toile** perdue
„	atum	toile à rendre
„	aturae	toile pour marchandise
„	aturi	plus tôt
„	aturos	le plus tôt possible
„	averam	ou plus **tôt**, si possible
„	averit	trop tôt
„	avero	il est encore trop tôt
„	avimus	télégraphiez le **total** (de)
„	avisse	à valoir sur le total
„	avisti	total jusqu'à ce jour
„	avit	par **train**
„	emus	„ „ partant
„	ent	**traite** due
„	es	traites dues
„	etis	traite pas arrivée
„	etur	traite en circulation
„	or	traites „ „

nec	abam	traite acceptée
„	abamus	retournez traite acceptée
„	abant	contre traite acceptée
„	abare	„ traites acceptées
„	abaris	traites acceptées
„	abatur	traite égarée
„	abimur	traites à payer
„	abis	télégraphiez si traite payée
„	abitis	traite a été payée
„	abo	„ n'a pas été payée
„	amini	si traite pas payée, télégraphiez
„	amur	payez traite
„	andam	payez traites
„	andi	payez traite à notre (ma) responsabilité
„	andior	payé traite à votre responsabilité
„	andius	devons-nous payer traite ?
„	andos	ne payez pas traite
„	ans	acceptons (accepte) traite
„	antem	acceptons (accepte) traite seulement pour
„	antis	acceptez traite (pour)
„	antium	en cas de non-acceptation de traite
„	anto	„ „ „ „ -payement „ „
„	antur	**transbordement** (à)
„	aremus	avec transbordement
„	arent	sans „
„	arere	télégraphiez à bord de quel vapeur marchandise a été transbordée
„	ares	aussitôt transbordé télégraphiez
„	aretis	évitez transbordement
„	aretur	**transfert** télégraphique
„	ari	transferts télégraphiques
„	arier	par transfert télégraphique
„	at	en **transit**
„	atae	droits de transit
„	atarum	frais de **transport**
„	atior	franco frais de transport
„	atius	pendant le transport
„	atos	**Tribunal** de Commerce
„	atote	beaucoup **trop**
„	atum	de trop
„	aturae	en trop
„	aturi	si trop
„	aturos	conforme au **type**
„	averam	inférieur „ „
„	averit	supérieur „ „
„	avero	envoyez type
„	avimus	„ types
„	avisse	envoyé type
„	avisti	„ types
„	avit	suivant type
„	emus	„ types
„	ent	l'**un** et l'autre
„	es	„ ou „
„	etis	„ et/ou l'autre
„	etur	à l'**unanimité**
„	or	très **urgent**

nigr	abam	l'affaire est urgente
„	abamus	l'affaire est excessivement urgente
„	abant	en cas d'urgence
„	abare	suivant les **usages** de la place
„	abaris	à l'**usine**
„	abatur	pris à l'usine
„	abimur	**valable** jusqu'à
„	abis	„ „ révocateon
„	abitis	„ pour 24 heures
„	abo	„ „ 2 jours
„	amini	„ „ 3 „
„	amur	„ „ 1 mois
„	andam	„ „ 2 „
„	andi	„ „ 3 „
„	andior	**valeur** approximative
„	andius	„ nominale
„	andos	„ totale
„	ans	„ déclarée
„	antem	„ à déclarer
„	antis	„ réalisable
„	antium	télégraphiez valeur (de)
„	anto	„ „ approximative
„	antur	„ „ totale
„	aremus	sur la valeur
„	arent	„ „ „ totale
„	arere	de la valeur
„	ares	contre-valeur
„	aretis	d'aucune valeur
„	aretur	à **valoir sur**
„	ari	plus-value
„	arier	moins-value
„	at	par **vapeur**
„	atac	„ „ ou voilier
„	atarum	vapeur direct
„	atior	par vapeur direct
„	atius	„ premier vapeur
„	atos	„ „ „ direct
„	atote	vapeur en charge (pour)
„	atum	par quel vapeur avez-vous expédié?
„	aturae	par quel vapeur expédierez-vous?
„	aturi	ex-vapeur
„	aturos	télégraphiez nom du vapeur
„	averam	il n'y a pas de vapeur en ce moment
„	averit	il y aura un autre vapeur
„	avero	ou autre vapeur
„	avimus	ou vapeur direct
„	avisse	à bord du vapeur
„	avisti	prochain vapeur partira
„	avit	quand partira prochain vapeur?
„	emus	tout **venant**
„	ent	nous venons (je viens) de
„	es	vient (viennent) d'être
„	etis	on vient de
„	etur	il y a eu des **ventes**
„	or	par suite de ventes

nod	abam	profitez de la première occasion pour vendre
„	abamus	ne forcez pas les ventes
„	abant	à valoir sur vente
„	abare	„ „ „ ventes
„	abaris	„ „ „ vente à découvert
„	abatur	„ „ „ ventes à „
„	abimur	impossible de vendre à aucun prix
„	abis	sauf vendu
„	abitis	c'est une vente d'occasion
„	abo	rien vendu
„	amini	espérons (espère) vendre plus tard à meilleur prix
„	amur	vendu sauf votre consentement
„	andam	peu de désir de vendre
„	andi	télégraphiez à qui vendu
„	andior	difficile à vendre
„	andius	presqu' invendable
„	andos	en cas de vente
„	ans	vendez pour notre compte
„	antem	„ „ mon „
„	antis	vendu pour votre compte
„	antium	on est en négociations pour la vente de
„	anto	nous sommes (je suis) en négociations pour la vente de
„	antur	il est plus avantageux de vendre que d'acheter
„	aremus	il est plus avantageux d'acheter que de vendre
„	arent	télegraphiez à quel prix vous pouvez vendre
„	arere	télégraphiez à quel prix vous recommandez de vendre
„	ares	télégraphiez quelle quantité et à quel prix vous pouvez vendre
„	aretis	télégraphiez quelle quantité et à quel prix vous recommandez de vendre
„	aretur	par suite de **vermoulure**
„	ari	marchandise attaquée par les vers
„	arier	bonification pour vermoulure
„	at	on exige bonification pour vermoulure
„	atae	après **vérification**
„	atarum	entièrement **versé**
„	atior	ou **vice-versa**
„	atius	et „ „
„	atos	en **vigueur**
„	atote	quand entrera en vigueur?
„	atum	on mettra en vigueur
„	aturae	plus en vigueur
„	aturi	pas encore en vigueur
„	aturos	modification entrera en vigueur
„	averam	le brevet a été **violé**
„	averit	**vis-à-vis** de
„	avero	grande **vitesse**
„	avimus	petite „
„	avisse	expédiez en grande vitesse
„	avisti	
„	avit	
„	emus	
„	ent	
„	es	
„	etis	
„	etur	
„	or	

oner	**abam**	expédiez en petite vitesse
„	**abamus**	expédié en grande „
„	**abant**	„ „ petite „
„	**abare**	sous **voiles**
„	**abaris**	par **voilier**
„	**abatur**	„ „ et vapeur
„	**abimur**	„ „ ou vapeur
„	**abis**	voilier direct
„	**abitis**	par voilier direct
„	**abo**	d'après ce que nous avons (j'ai) **vu** (de)
„	**amini**	„ „ qu'on **a** vu (de)
„	**amur**	en **vrac**
„	**andam**	expédiez en vrac
„	**andi**	expédié „ „
„	**andior**	est-il **vrai** que?
„	**andius**	télégraphiez s'il est vrai
„	**andos**	il est vrai (que)
„	**ans**	il n'est pas vrai (que)
„	**antem**	s'il est vrai (que)
„	**antis**	**vu** que
„	**antium**	en vue de
„	**anto**	à vue
„	**antur**	„ 8 jours de vue
„	**aremus**	„ 10 „ „ „
„	**arent**	„ 15 „ „ „
„	**arere**	„ 30 „ „ „
„	**ares**	„ 45 „ „ „
„	**aretis**	„ 60 „ „ „
„	**aretur**	„ 3 mois „ „
„	**ari**	„ 4 „ „ „
„	**arier**	„ 6 „ „ „
„	**at**	sur **wagon**
„	**atae**	rendu sur wagon
„	**atarum**	„ „ „ votre place
„	**atior**	„ „ „ notre place
„	**atius**	par wagon
„	**atos**	wagon complet
„	**atote**	wagon de 5000 Ko
„	**atum**	„ „ 10000 „
„	**aturae**	„ „ 15000 „
„	**aturi**	par wagon de 5000 Ko
„	**aturos**	„ „ „ 10000 „
„	**averam**	expédiez 1 wagon de 10000 Ko
„	**averit**	„ 2 wagons „ „ „
„	**avero**	à **vous**
„	**avimus**	de vous
„	**avisse**	par vous
„	**avisti**	pour vous
„	**avit**	sur vous
„	**emus**	il **y** a
„	**ent**	il n'y a pas
„	**es**	s'il y a
„	**etis**	s'il n'y a pas
„	**etur**	télégraphiez s'il y a
„	**or**	

Table 1.

58 roots ×

Correspondence and Telegrams.

		your telegram of / *votre télégramme du*	your letter of / *votre lettre du*
		→	→
as per	*suivant*	opac–	refl–
as per . . . . . we (I) **accept**	*suivant . . . . nous (j') **accept**/ons(e)*	opim–	regn–
we (I) accept offer made in	*acceptons (j'accepte) offre faite dans*	opt–	reme–
impossible to accept offer made in	*impossible d'accepter offre faite dans*	orn–	rog–
we (I) **agree** with	*nous sommes (je suis) d'**accord** avec*	palp–	ruct–
act as proposed in	***agissez** comme proposé dans*	pand–	sacc–
shall act as proposed in	*agirons (agirai) comme proposé dans*	pari–	salv–
impossible to act as proposed in	*impossible d'agir comme proposé dans*	patr–	sati–
we (I) recommend not to act as proposed in	*nous (je) recommand/ons(e) de ne pas agir comme proposé dans*	ped–	serv–
we (I) have acted as proposed in	*avons (ai) agi comme proposé dans*	plag–	sim–
we (I) shall act as mentioned in	*agirons (agirai) comme mentionné dans*	plic–	sinu–
impossible to act as mentioned in	*impossible d'agir comme mentionné dans*	plor–	soci–
we (I) have acted as mentioned in	*avons (ai) agi comme mentionné dans*	plum–	sper–
we (I) leave you to act for the best, this is in reply to	*faites au mieux; ceci en réponse à*	port–	spic–
there must be a **mistake** in	*il doit y avoir **erreur** dans*	priv–	spum–
we (I) **received**	*nous avons (j'ai) **reçu***	prob–	stip–
we (I) did not receive	*nous n'avons (je n'ai) pas reçu*	puls–	sud–
we (I) **reply** in the affirmative to	*nous (je) **répond**/ons(s) affirmativement à*	purg–	sulc–
we (I) reply in the negative to	*nous (je) répond/ons(s) négativement à*	put–	tali–
ship (send) as per	***expédiez** suivant*	radi–	tard–
we (I) have shipped (sent) as per	*nous avons (j'ai) expédié suivant*	rapt–	tax–

		our (my) telegram of / ***notre** (mon) télégramme du*	our (my) letter of / ***notre** (ma) lettre du*
		→	→
as per	*suivant*	tenu–	vast–
please **reply** to	*veuillez **répondre** à*	torp–	vect–
did you **receive**?	*avez-vous **reçu?***	tran–	vex–
we (I) **confirm**	*nous **confirm**ons (je confirme)*	uber–	vibr–
there is a mistake, **correct** as follows	*il y a une erreur, rectifiez comme suit*	ulul–	viol–
when did you **send** goods ordered by?	*quand avez-vous **expédié** marchandise(s) commandée(s) par?*	umbr–	viti–
when are you sending goods ordered by?	*quand expédierez-vous marchandise(s) commandée(s) par?*	vacu–	volg–
we (I) renew for immediate telegraphic reply **offer** made by	*nous renouvelons (je renouvelle) pour réponse télégraphique immédiate, **offre** faite par*	vall–	vor–

55 **terminations** = 3190 ciphers (Latin). **Table 1.**

	of **current** month *du mois **courant***	of **last** month *du mois **passé***
	←	←
1 st / er	-abam	—
2 nd / —	-abamus	—
3 rd / —	-abant	—
4 th / —	-abare	—
5 „	-abaris	—
6 „	-abatur	—
7 „	-abimur	—
8 „	-abis	-at
9 „	-abitis	-atae
10 „	-abo	-atarum
11 „	-amini	-atior
12 „	-amur	-atius
13 „	-andam	-atos
14 „	-andi	-atote
15 „	-andior	-atum
16 „	-andius	-aturae
17 „	-andos	-aturi
18 „	-ans	-aturos
19 „	-antem	-averam
20 „	-antis	-averit
21 st / —	-antium	-avero
22 nd / —	-anto	-avimus
23 rd / —	-antur	-avisse
24 th / —	-aremus	-avisti
25 „	-arent	-avit
26 „	-arere	-emus
27 „	-ares	-ent
28 „	-aretis	-es
29 „	-aretur	-etis
30 „	-ari	-etur
31 st / —	-arier	-or

Tables:
Co & tel.
1

Examples:

opimabant	=	as per your tel. of 3rd of curr. month we (I) accept. *suivant r. tél. du 3 ct. nous acceptons (j'accepte).*
sudatum	=	we (I) did not receive your letter of 15th last month. *nous n'avons (je n'ai) pas reçu votre lettre du 15 p.*
vallor	≐	we (I) ren. f. i. t. r. offer m. by our (my) tel. 31st last m. *n. ren. (je r.) p. r. t. i. offre f. p. notre (mon) tél. du 31 m. passé.*

acer-	re the	*concernant le*
acun-	wire **reply** to our (my) last telegram re the	*télégraphiez* ***réponse*** *à notre (mon) dernier télégramme concernant le*
adat-	we (I) have understood the .... of your telegram as follows, if correct, need not reply	*nous avons (j'ai) entendu le ...... de votre télégramme comme suit, si correct, pas nécessaire répondre*
adoj-	as per your last telegram we (I) accept **offer** expressed by the	*suivant votre dernier télégramme nous acceptons (j'accepte)* ***offre*** *exprimée par le*
adub-	as per your telegram preceding your last telegram, we (I) accept **offer** expressed by the	*suivant votre avant-dernier télégramme nous acceptons (j'accepte)* ***offre*** *exprimée par le*
aerm-	as to your telegram wire **more clearly** your meaning expressed by the	*quant à votre télégramme télégraphiez* ***plus clairement*** *ce que vous voulez dire par le*

	telegram the *télégramme le*	yesterday's telegram the *télégramme d'hier le*	**first** telegram of to-day the ***1er*** *télégramme de ce jour le*	**2nd** telegram of to-day the ***2e*** *télégramme de ce jour le*	**last** telegram sent the ***dernier*** *télégramme envoyé le*
repeat of your *répétez de votre*	afam-	agol-	ahit-	alij-	anul-
we (I) **repeat as** follows of our (my) *nous* ***répétons*** *(je répète) comme suit de notre (mon)*	afil-	agri-	ahoy-	alom-	apup-
there must be a **mistake**, examine and correct, if necessary, of your *il doit y avoir* ***erreur,*** *examinez et, si nécessaire, rectifiez de votre*	afod-	aguj-	ajeb-	anot-	aruf-

Examples:

adojabala = as p. y. l. tel. we (I) acc. offer expr. by the 2nd wd.
s. v. d. tél. nous acceptons (j'acc.) offre e. p. le 2e mot

agujadnos = th. m. be a mistake. ex. and corr. if nec. of yr. yest. tel. the 12th wd.
il d. y av. erreur ex. et, si n., rectifiez d. v. tél. d'hier le 12e mot.

56 **terminations** = 1176 ciphers (Spanish). **Table 2.**

	word (cipher) *mot (chiffre)*	word (cipher) and the word (cipher) following it *mot (chiffre) et le mot (chiffre) qui le suit*	word (cipher) and the 2 words (ciphers) following it *mot (chiffre) et les 2 mots (chiffres) qui le suivent*
	←	←	←
1st / 1er	-aba	-andote	-arieis
2nd / 2e	-abala	-anlas	-arlos
3rd / 3e	-abales	-anlo	-aronlo
4th / 4e	-abame	-anme	-aronte
5 „	-abamos	-arais	-arse
6 „	-abanlo	-aralas	-aseis
7 „	-abaste	-aralo	-astelo
8 „	-ad	-aranla	-elas
9 „	-adas	-aranme	-elo
10 „	-adla	-aranos	-enla
11 „	-adles	-arate	-enles
12 „	-adnos	-arela	-enos
13 „	-adolas	-areles	-ense
14 „	-adolo	-areme	-ete
15 „	-adome	-aremos	-ola
16 „	-ale	-aren	-oles
17 „	-amonos	-ariala	-omelos
18 „	-amos	-ariame	-ose
19 „	-andola	-arian	—

Table 3 Dates.

72 roots ×

as per your **letter** dated: *suivant votre lettre datée:*	as per our (my) **letter** dated: *suivant notre (ma) lettre datée:*	as per Invoice dated: *suivant facture datée:*	**shipment** (dispatch) *expédition*	**due** *dû*	~~~~~		
→	→	→	→	→	→		
abem–	clep–	emet–	futu–	mitt–	pon–	January	*Janvier*
acu–	clod–	emol–	gign–	nasc–	posc–	February	*Février*
aded–	coal–	erep–	glub–	nect–	prem–	March	*Mars*
adig–	coem–	erig–	imbu–	nub–	reag–	April	*Avril*
adur–	coqu–	erod–	inig–	obed–	redd–	May	*Mai*
allu–	deps–	exed–	intu–	obsu–	renu–	June	*Juin*
argu–	deru–	exim–	inur–	pang–	rump–	July	*Juillet*
aveh–	dilu–	exur–	irru–	parc–	scab–	August	*Août*
bib–	disc–	find–	juss–	pell–	sequ–	September	*Septembre*
caed–	edic–	flig–	lamb–	pend–	serp–	October	*Octobre*
cill–	eger–	frum–	leg–	pet–	sist–	November	*Novembre*
cing–	elud–	fung–	metu–	pins–	.solv–	December	*Décembre*

Table 4 Dates.

34 roots ×

→		
alap–	**about**	***environ***
amar–	**arrival**	***arrivée***
amic–	probable arrival on	„ *probable le*
anax–	arrival not later than	„ *pas plus tard que le*
artu–	to **arrive** in ballast on	*à arriver sur son leste le*
baub–	**before**	***avant*** *le*
bell–	not before	*pas avant le*
cert–	transaction **cancelled** unless arrived on	*transaction* ***annulée*** *à moins d'arrivée le*
clin–	cancelling date	*date d'annulation*
dupl–	**delivery**	***livraison***
egel–	extension until	***prorogation*** *jusqu'au*
epot–	„ granted until	„ *accordée jusqu'au*
erug–	in **force** on and after	*en* ***vigueur*** *le et dès le*
evag–	not **later** than	*pas plus* ***tard*** *que le*
flex–	**departure**	***départ***
flut–	probable departure	„ *probable*
idu–	{firm **offer** for reply to reach before 12 o' clock on	{***offre*** *ferme pour réponse à parvenir avant midi le*
just–	{firm offer for reply to reach before 4 p. m. on	{*offre ferme pour réponse à parvenir avant 4 h. du soir le*
maci–	endeavour to extend offer until	*tâchez de proroger offre jusqu'au*
mir–	we (I) extend firm offer until	{*nous prorogeons (je proroge) offre ferme jusqu'au*
mot–	vessel **ready** to load on	*bateau* ***prêt*** *à charger le*
muss–	„ „ „ unload on	„ „ „ *décharger le*
obar–	the **remainder** (balance)	*le* ***reste*** *(le solde)*
obes–	we (I) shall **send** (ship) on	*nous* ***expédierons*** *(j'expédierai) le*
ocul–	goods ordered will be sent (shipped) on	*marchandise commandée s'expédiera le*
olet–	to be sent (shipped) on	*à expédier le*
omin–	„ „ „ „ latest on	„ „ *au plus tard le*
ovic–	impossible to send (ship) before	*impossible d'expédier avant le*
pis–	**until**	***jusqu'au***
pomp–	1/4	1/4
rauc–	1/3	1/3
ret–	1/2	1/2
runc–	2/3	2/3
rusp–	3/4	3/4

31 terminations = 2232 ciphers (Latin) **Table 3.**

-am	1	st / er
-amus	2	nd
-ant	3	rd
-atis	4	th
-atur	5	„
-ebamus	6	„
-ebant	7	„
-ebare	8	„
-ebas	9	„
-ebatis	10	„
-ebatur	11	„
-emini	12	th
-endae	13	„
-endi	14	„
-endior	15	„
-endius	16	„
-endos	17	„
-ens	18	„
-entem	19	„
-entis	20	„
-entium	21	st
-entur	22	nd
-erem	23	rd
-eremus	24	th
-erent	25	„
-erere	26	„
-eretis	27	„
-eretur	28	„
-eris	29	„
-et	30	„
-imur	31	st

Examples:

glubatur = shipment (dispatch) 5th March. — *expédition 5 Mars.*
ponebas = 9th January. — *9 Janvier.*

39 terminations = 1326 ciphers (Latin) **Table 4.**

-abam	1	st / er	of current month — *du mois* ***courant***
-abamus	2	nd	
-abant	3	rd	
-abare	4	th	
-abaris	5	„	
-abatur	6	„	
-abimur	7	„	
-abis	8	„	
-abitis	9	„	
-abo	10	„	
-amini	11	„	
-amur	12	„	
-andam	13	„	
-andi	14	„	
-andior	15	„	
-andius	16	„	
-andos	17	„	
-ans	18	„	
-antem	19	„	
-antis	20	„	
-antium	21	st	
-anto	22	nd	
-antur	23	rd	
-aremus	24	th	
-arent	25	„	
-arere	26	„	
-ares	27	„	
-aretis	28	„	
-aretur	29	„	
-ari	30	„	
-arier	31	st	

-at	1	st / er	of next month — *du mois* ***prochain***
-atote	2	nd	
-emus	3	rd	
-ent	4	th	
-es	5	„	
-etis	6	„	
-etur	15	„	
-or	ultimo		

Dates &c. 3

Examples:

epotetur = extension granted until 15th next m. — *prorogat. acc. jusqu'au 15 mois proch.*
flutantis = prob. departure 20th curr. m. — *départ probable le 20 mois cour.*

Table 5 Epochs *Epoques* 14 roots ×

→	~~~~	~~~~
abus–		
acet–	about	*environ*
acqu–	before	*avant*
adir–	not before	*pas avant*
adon–	delivery	*livraison*
adul–	deliveries	*livraisons*
agun–	during	*pendant*
ajut–	for	*pour*
ambr–	from	*dès*
amen–	not later than	*pas plus tard que*
asol–	probably	*probablement*
attu–	shipment (dispatch)	*expédition*
bagn–	shipments	*expéditions*
barr–	until	*jusqu'à*

Examples:

adirammolo = not before beginning of next month — *pas avant comm. m. proch.*
abusiamo = next week — *semaine prochaine.*
ameniamo = not later than next week — *pas plus tard que semaine prochaine.*

Table 6 Epochs *Epoques* 48 roots ×

~~	delivery *livraison*	shipment (dispatch) *expédition*	firm offer for immediate telegraph. reply for shipment (dispatch) *offre ferme pour réponse télégraph. immédiate pour expédition*		
→	→	→	→		
absum–	arrod–	decid-	efflu–	January	*Janvier*
accin–	asser–	deflu–	eling–	February	*Février*
accud–	atter–	degun–	emerg–	March	*Mars*
adbib–	attig–	delin–	emung–	April	*Avril*
adduc–	avell–	depon–	evolv–	May	*Mai*
adind–	averr–	depuv–	excud–	June	*Juin*
adrad–	battu–	derad–	exist–	July	*Juillet*
allid–	coadd–	desug–	exlid–	August	*Août*
ambed–	comed–	deteg–	expet–	September	*Septembre*
ambig–	consu–	deung–	expon–	October	*Octobre*
ambur–	corru–	dirim–	exser–	November	*Novembre*
arrig–	decan–	divid–	exsug–	December	*Décembre*

Examples:

arrodito = delivery January. — *livraison Janvier.*
emunget = { f. o. f. i. t. r. f. shipment April in equ. monthly lots to end of Oct.
offre f. p. r. t. i. p. exp. Avril en lots ég. mensuels jusqu'à fin Oct. }

83 terminations = 1162 ciphers (Italian) **Table 5.**

beginning of *commencement de*	middle of *mi*	end of *fin*	~~~~~~		
←	←	←	←		
-acelo	-assero	-ava	-erete	last week	*semaine passée*
-acene	-assimo	-avaci	-iamla	current „	*„ courante*
-aile	-astene	-avala	-iamo	next „	*„ prochaine*
-aivi	-astevi	-avamo	-iamoci	{the week after {next week	*{ „ après la* *{semaine prochaine*
-ali	-asti	-avanlo	-iamolo	last month	*mois passé*
-ammo	-astici	-avate	-iamone	current month	*„ courant*
-ammolo	-astila	-aveli	-iate	next „	*„ prochain*
-ammone	-atale	-avilo	-icela	{the month after {next month	*{mois après* *{le mois prochain*
-ammovi	-atasi	-avimi	-ila	last year	*l'année passée*
-ando	-atela	-avine	-imelo	present „	*„ courante*
-andoci	-atemi	-avole	-imene	next „	*„ prochaine*
-andole	-atili	-avoti	-imi	last season	*saison passée*
-ansene	-ato	-era	-ino	present „	*„ courante*
-ansi	—	—	-oceli	next „	*„ prochaine*
-ante	—	-eranno	-ole	last quarter	*trimestre passé*
-antesi	-atolo	-erebbe	-olla	present „	*„ en cours*
-arci	—	—	-omele	next „	*„ prochain*
-are	-atone	-erei	-ommi	this spring	*ce printemps*
-arla	—	—	-onne	next „	*printemps prochain*
—	-atovi	-eremo	-ossi	this summer	*cet été*
-armene	—	—	-otelo	next „	*été prochain*
-aronle	—	—	-otti	this autumn	*cet automne*
-arono	—	—	-ovela	next „	*automne prochain*
-aronsi	—	-eresti	-ovene	this winter	*cet hiver*
-asse	—	—	-ovi	next „	*hiver prochain*

25 terminations = 1200 ciphers (Latin) **Table 6.**

to end of *jusqu'à fin*	in equal monthly lots to end of *en lots égaux mensuels* *jusqu'à fin*		
←	←		
-am	-endum	January	*Janvier*
-amus	-ens	February	*Février*
-ant	-entem	March	*Mars*
-atis	-entis	April	*Avril*
-atur	-entur	May	*Mai*
-ebant	-erem	June	*Juin*
-ebare	-erent	July	*Juillet*
-ebas	-erere	August	*Août*
-emini	-eris	September	*Septembre*
-endae	-et	October	*Octobre*
-endi	-imur	November	*Novembre*
-endos	-is	December	*Décembre*

←

-ito ~~~~~~

Table 7 Hours *(Heures)*. 38 roots ×

		yesterday *hier*	to-day *aujourd'hui*	to-morrow *demain*	day after to-morrow *après-demain*
		longi–	nitid–	recol–	spons–
we (I) shall arrive	*nous(j')***arriver***|ons(ai)*	—	obhum–	redam–	stell–
will arrive	*arriver|ont(a)*	—	obsip–	redon–	stigm–
arrived	*arrivé*	maest–	obunc–	—	—
we (I) shall leave	*nous (je)* **partir***|ons(ai)*	—	obvel–	refug–	subar–
will leave	*partir|ont(a)*	—	opert–	reorn–	sucid–
left	*parti*	negum–	oppic–	—	—
firm offer for reply to reach latest	***offre*** *ferme pour réponse à parvenir au plus tard*	—	opput–	rigid–	tabul–
against your offer we (I) make you counter-offer for reply to reach latest	*contre votre offre nous vous faisons (je vous fais) contre-offre pour réponse à parvenir au plus tard*	—	oscit–	risit–	tamin–
we (I) renew firm offer for reply to reach latest	*nous renouvelons (je renouvelle) offre ferme pour réponse à parvenir au plus tard*	—	pavid–	scrut–	tepid–
reply to reach latest	***réponse*** *à parvenir au plus tard*	—	pejor–	sejug–	tripl–
affirmative reply to reach latest	*réponse affirmative à parvenir au plus tard*	—	press–	sibil–	tunic–
negative reply to reach latest	*réponse négative à parvenir au plus tard*	—	racem–	simpl–	velit–

Table 8 Hours *(Heures)*. 24 roots ×

a. m. *matin*	p. m. *soir*	
pront–	rilut–	1,
quart–	rimen–	2,
regal–	rimur–	3,
rialz–	rinom–	4,
riatt–	riorn–	5,
ricap–	ripos–	6,
ridam–	risan–	7,
ridon–	rivel–	8,
rifor–	rotol–	9,
rigir–	rovin–	10,
rilim–	rucol–	11,
rilod–	rugum–	12,

Examples:

rimenali = 2,04 p. m. — 2,04 *soir.*

ricaparono = 6,11 a. m. — 6,11 *matin.*

28 **terminations** = 1064 ciphers (Latin) **Table 7.**

	a. m. *matin*	p. m *soir*
1	-abam	-antem
2	-abant	-antis
3	-abare	-anto
4	-abis	-antur
5	-abo	-arent
6	-amini	-arere
7	-amur	-ares
8	-andae	-ari
9	-andi	-arier
10	-andos	-at
11	-andum	-atote
12	-ans	-emus

-ent	morning	*matin*
-es	afternoon	*après-midi*
-etis	evening	*soir*
-etur		

Examples:

redonamur = will arrive to-morr. 7 a. m. — *arriver/ont(a) dem. 7 h. mat.*
longietis = yesterday evening — *hier soir*
negumetur = left yesterday — *parti hier*

60 **terminations** = 1440 ciphers (Italian) **Table 8.**

-acelo	00	-ato	20	-iamo	40
-acene	01	-atolo	21	-iate	41
-aile	02	-atone	22	-icela	42
-aivi	03	-atovi	23	-ila	43
-ali	04	-ava	24	-imelo	44
-ammo	05	-avaci	25	-imene	45
-ansi	06	-avala	26	-imi	46
-ante	07	-avamo	27	-ino	47
-arci	08	-avate	28	-iteli	48
-are	09	-aveli	29	-oceli	49
-arla	10	-avilo	30	-ole	50
-arono	11	-avimi	31	-olla	51
-asele	12	-avine	32	-omele	52
-asse	13	-avole	33	-ommi	53
-asti	14	-avoti	34	-ossi	54
-atale	15	-era	35	-otelo	55
-atasi	16	-erei	36	-otti	56
-atela	17	-eremo	37	-ovela	57
-atemi	18	-erete	38	-ovene	58
-atili	19	-iamla	39	-ovi	59

Table 9.

88 roots X

Asking for Information. *Demandes de Renseignements.*

Inquiries are understood for information under discretion and without guarantee.

Demandes s'entendent pour renseignements discrets et sans garantie.

		of your town *de votre ville*	~~~~~
~~	~~~~~	➜	➜
.... &	*.... & ....*	demet-	habit-
Brothers	*frères*	deneg-	hebet-
„ & Co.	*„ & Co.*	denot-	hilar-
& Brothers	*& frères*	depil-	ignor-
		deput-	illim-
Company	*Compagnie (de(s)*	deriv-	immol-
& Co.	*& Co.*	derog-	immut-
Son	*fils*	desol-	imper-
„ & Co.	*„ & Co.*	desud-	incer-
& Son	*& fils*	devit-	indag-
& Sons	*& fils*	devor-	indic-
Sons	*fils*	diffl-	indur-
„ & Co.	*„ & Co.*	dijug-	inesc-
Father & Son	*père & fils*	dilat-	infam-
Ltd	***Ltd***	edent-	innod-
Brothers Ltd		effug-	inocc-
Brothers & Co Ltd		elogi-	insol-
& Brothers Ltd		emend-	intim-
Company Ltd		emigr-	inunc-
& Co. Ltd		enerv-	invit-
Son Ltd		equit-	invoc-
„ & Co. Ltd		evulg-	irrig-
& Son Ltd		excav-	irruf-
Sons Ltd		excit-	judic-
„ & Co. Ltd		excus-	labor-
& Sons Ltd		exhal-	lacun-
Société Anonyme		exopt-	lapid-
frères Socté An.		exoss-	levig-
& frères „ „		expil-	loric-
& Co. „ „		exput-	lumin-
fils „ „		exsol-	lustr-
„ & Co. „ „		exsud-	lymph-
& fils „ „		exund-	macer-
Société en Commandite		famul-	madid-
Aktien-Gesellschaft		febru-	malax-
Gebrüder Act.-Ges.		figul-	marit-
& Co. „ „		flagr-	matur-
Sohn „ „		flamm-	memor-
„ & Co. „ „		focil-	minor-
& Sohn „ „		fraud-	mitig-
Söhne „ „		fugit-	moder-
„ & Co. „ „		funer-	numer-
& Söhne „ „		gener-	nunti-
Ges. m. b. H.		glaci-	obdur-

Examples:

Meier excusamini = { Meier Sons & Co. Ltd of your town wire inform., part. ab. compet. — *„ Sons & Co. Ltd de v. v. télégr. rens. surt. conc. compét.*

Girard indagandos Odessa = { Girard & Son of Odessa wire inf. & wh. m. be cons. good f. cred. of some hundred doll. — *Girard & fils d'Odessa tél. rens. & si c. m. p. être cons. bonne p. créd. de quelques cent. de doll.*

King Goldstone denegatior = { King & Goldstone of your town wire inf.; this inqu. is caus. by rum. of imp. changes in the firm. — *King & Goldstone de v. v. tél. rens.; cette dem. est c. p. rum. de chang. imm. dans la m.*

-abam	～～～～～～	～～～～～～
-abant	send information by letter (ordinary report)	*envoyez renseignements par lettre (rapport ordinaire)*
-abare	send **full** information (special report) by letter	*envoyez renseignements détaillés (rapport spécial) par lettre*
-abis	wire information and send **full** (special) report by letter	*télégraphiez renseignements et envoyez rapport détaillé (spécial) par lettre*
-abo	wire information, particularly about	*télégraphiez renseignements, surtout concernant*
-amini	wire information, particularly about competence	*télégraphiez renseignements, surtout concernant compétence*
-amur	wire information, particularly about means	*télégraphiez renseignements, surtout concernant moyens*

	wire information and whether might be considered good for a	*télégraphiez renseignements et si cette maison (personne) pourrait être considérée bonne pour un*
-andae	**credit**	***crédit***
-andi	„ of some hundred £	„ *de quelques centaines de* £
-andos	„ „ „ „ dollars	„ „ „ „ „ *dollars*
-andum	„ „ „ „ francs	„ „ „ „ „ *francs*
-ans	„ „ „ thousand £	„ „ „ *milliers* „ £
-antem	„ „ „ „ dollars	„ „ „ „ „ *dollars*
-antis	„ „ „ „ francs	„ „ „ „ „ *francs*

	wire information; this inquiry is caused:	*télégraphiez renseignements; cette demande est* ***causée****:*
-anto	by	*par*
-antur	„ apparent financial difficulties	„ *difficultés apparentes financières*
-arent	„ draft returned unpaid	„ *traite retournée impayée*
-arere	„ reports of losses	„ *rapports de pertes*
-ares	„ reports of losses by failure (of)	„ *rapports de pertes par faillite (de)*
-ari	„ reports of losses by speculation	„ *rapports de pertes par spéculation*
-arier	„ reports of failure	„ *rapports de faillite*
-at	„ disquieting rumours	„ *rumeurs inquiétantes*
-atae	„ rumours of impending failure	„ *rumeurs de faillite imminente*
-atior	„ by rumours of impending changes in the firm	„ *rumeurs de changements imminents dans la maison*

Inf. & Fin 9

	wire information; this firm (person) is established as:	*télégraphiez renseignements; cette maison (personne) est* ***établie*** *comme:*
-atius	agent(s)	*agent(s)*
-atos	banker(s)	*banquier(s)*
-atote	broker(s)	*courtier(s)*
-atum	exporter(s)	*exportateur(s)*
-aturi	„ and importer(s)	„ *et importateur(s)*
-avero	finance broker(s)	*courtier(s) en finance*
-avit	forwarding agent(s)	*expéditeur(s)*
-emus	importer(s)	*importateur(s)*
-ent	manufacturer(s)	*fabricant(s)*
-es	retail firm	*maison de détail*
-etis	shipbroker(s)	*courtier(s) de navire*
-etur	wholesale firm	*maison de gros*
-or	„ and retail firm	„ „ „ *et détail*

Table 10.

133 roots ×

Giving Information *Renseignements fournis*

information under discretion and without guarantee — *renseignements discrets et sans garantie.*

		the firm is very large *la maison est très grande*	the firm is large *la maison est grande*	the firm is of medium importance *la maison est d'importance moyenne*	the firm is small *la maison est petite*	is said to be: *on dit qu'elle:* very competent *est très compétente*	competent *est compétente*	not sufficiently competent *n'est pas suffisamment compétente*
the firm (person) in question is established as:	***maison (personne) en question est établie comme:***	➞	➞	➞	➞	➞	➞	➞
agent(s)	*agent(s)*	abbon-	adast-	aggel-	annus-	asper-	blind-	compr-
forwarding agent(s)	*expéditeur(s)*	abbuj-	addit-	—	—	assit-	bulin-	comun-
banker(s)	*banquier(s)*	abiss-	addom-	aggir-	—	asson-	capon-	coort-
broker(s)	*courtier(s)*	abjur-	adegu-	agogn-	appaj-	attel-	carat-	covid-
finance broker(s)	„ *en finance*	abond-	adizz-	alber-	—	attiv-	caren-	crost-
ship „	„ *de navire*	abort-	adjuv-	—	—	avall-	censu-	cucin-
exporter(s)	*exportateur(s)*	abram-	adorb-	allir-	arbor-	avanz-	chiod-	decim-
„ and importer(s)	„ *et importateur(s)*	accas-	adugn-	amatt-	argan-	avver-	cicur-	decor-
importer(s)	*importateur(s)*	accod-	affam-	ammaj-	arnes-	avvil-	ciuff-	delib-
manufacturer(s)	*fabricant(s)*	accul-	affid-	ammir-	arred-	azzim-	ciurm-	denod-
retail firm	*maison de détail*	acerb-	affun-	ammod-	arriv-	bajul-	class-	depur-
wholesale firm	„ „ *gros*	acesm-	agazz-	anell-	arrol-	biond-	coltr-	desal-
„ and retail firm	„ „ „ *et détail*	adaqu-	agenz-	annev-	arrub-	bitum-	comod-	desin-

		morality is said to be *on dit que **moralité** est* excellent *excellente*	very good *très bonne*	good *bonne*	fairly good *assez bonne*	moderate *modérée*	bad *mauvaise*	morality: nothing unfavorable is known ***moralité:** rien de défavorable n'est connu*	~~~~~~
the competence of the firm (person) in question is said to be:	*on dit que la **compétence** de la maison (personne) en question est:*	➞	➞	➞	➞	➞	➞	➞	➞
excellent	*excellente*	deton-	dipan-	disat-	eccit-	etern-	—	grass-	—
very good	*très bonne*	difil-	dipel-	disir-	enarr-	feltr-	fromb-	gratt-	imbas-
good	*bonne*	dimoj-	diput-	disus-	erutt-	figur-	frust-	grond-	imbol-
fairly good	*assez bonne*	dinot-	diren-	divin-	esecr-	foder-	fucil-	grupp-	immel-
moderate	*modérée*	dinud-	dirob-	divor-	esorn-	frapp-	giamb-	guazz-	immit-
insufficient	*insuffisante*	—	dirup-	duell-	espir-	frett-	giunt-	illum-	immor-

➞

impal-	it is said that	*on dit que*
impep-	according to general opinion	*suivant l'opinion générale*
impur-	~~~~~~	~~~~~~

Table 10.

might be considered good for credit: *pourrait être considérée bonne pour crédit:*	might be considered good against acceptance of draft for credit: *pourrait être considérée bonne contre acceptation de traite pour crédit:*		
-acelo	—	of considerable extent	*considérable*
-acene	-arono	„ ordinary „	*ordinaire*
-aile	-asele	„ moderate „	*modéré*
-aivi	-asse	„ restricted „	*d'étendue restreinte*
-ali	-asti	to the extent of its (his) trade requirements	*proportionné aux exigences de ses affaires*
-ammo	-atale	to the extent mentioned	*à l'étendue mentionnée*
-ando	-atasi	of some hundred £	*de quelques centaines de £*
-ansi	-atela	„ „ „ dollars	*„ „ „ de dollars*
-ante	-atemi	„ „ „ francs	*„ „ „ „ francs*
-arci	-atili	„ „ thousand £	*„ „ milliers de £*
-are	-ato	„ „ „ dollars	*„ „ „ „ dollars*
-arla	-atolo	„ „ „ francs	*„ „ „ „ francs*

means are *moyens sont*	means apparently *moyens paraissent*				
-atone	—	enormous		*énormes*	
-atovi	—	very considerable		*très considérables*	
-ava	-iamla	considerable		*considérables*	
-avaci	-iamo	fairly considerable		*assez considérables*	
-avala	-iate	moderate		*modérés*	
-avamo	-icela	small		*petits*	
-avate	-ila	very small		*très petits*	
-aveli	-imelo	nil		*nuls*	
-avilo	-imene	sufficient	for its (his) transactions	*suffisants*	*pour ses transactions*
-avimi	-imi	hardly sufficient	for its (his) transactions	*à peine suffisants*	*pour ses transactions*
-avine	-ino	insufficient	for its (his) transactions	*insuffisants*	*pour ses transactions*
-avole	—	small	credit does not seem advisable unless against guarantee	*petits*	*crédit ne semble pas à propos sinon contre garantie*
-avoti	—	very small	credit does not seem advisable unless against guarantee	*très petits*	*crédit ne semble pas à propos sinon contre garantie*
-era	-iteli	hardly sufficient for its (his) transactions	credit does not seem advisable unless against guarantee	*à peine suffisants p. ses transactions*	*crédit ne semble pas à propos sinon contre garantie*
-erei	-oceli	insufficient for its (his) transactions	credit does not seem advisable unless against guarantee	*insuffisants pour ses transactions*	*crédit ne semble pas à propos sinon contre garantie*
-eremo	-ole	nil	it seems advisable not to grant any credit	*nuls*	*il semble à propos de n'accorder aucun crédit*
-erete	-olla	insufficient for its (his) transactions	it seems advisable not to grant any credit	*insuffisants pour ses transactions*	*il semble à propos de n'accorder aucun crédit*

-omele	is not considered good for any credit	*n'est considérée bonne pour aucun crédit*
-ommi	not known to have means	*on ne lui connait pas de moyens*
-onne	financial position is not clear	*la situation financière n'est pas claire*
-ossi	„ „ „ critical	*„ „ „ est critique*
-otelo	means inascertainable	*moyens inconstatables*
-otti	nothing unfavorable is known about payments	*rien de défavorable n'est connu concernant paiements*
-ovela	is a slow payer	*paie lentement*
-ovene	always paid regularly	*toujours payé régulièrement*
-ovi		

Table 11 Finance.

57 roots ×

		~~	balance *solde*	Invoice amount *montant de facture*
draw on us (me)	***fournissez*** *sur nous (moi)*	flect-	insum-	obrod-
may we (I) draw on you?	*pouvons-nous (puis-je) fournir sur vous?*	frang-	integ-	obteg-
shall draw on you unless you wire to the contrary	*fournir/ons(ai) sur vous sauf avis contraire par télégr.*	illin-	inung-	occid-
we (I) have drawn on you	*avons(ai) fourni sur vous*	illud-	invad-	occul-
do not draw, shall remit	*ne fournissez pas, remettr/ons(ai)*	imbib-	inveh-	offig-
do not draw, have remitted	*ne fournissez pas, avons (ai) remis*	impet-	irrad-	oppet-
remit	***remettez***	impon-	irrep-	oppon-
shall remit	*remettr/ons(ai)*	incol-	obbib-	perag-
we (I) have remitted	*avons (ai) remis*	incud-	obduc-	pered-
do not remit	*ne remettez pas*	induc-	obger-	perur-
do not remit, shall draw	*ne remettez pas, fournir/ons(ai)*	inger-	oblid-	proru-
„ „ „ we (I) have drawn	*„ „ „ avons(ai) fourni*	insip-	obnub-	quaer-

recog-	exchange	*change*
refig-	wire exchange	*télégraphiez change*
releg-	draft	*traite*
relid-	drafts	*traites*
repet-	remittances	*remises*
resum-	against remittance	*contre remise*
retex-	„ draft	*„ traite*
reveh-	„ „ accepted	*„ traite acceptée*
revom-	1/3	1/3
scind-	2/3	2/3
scisc-	1/2	1/2
scrib-	balance	*solde*
sculp-	it is more advantageous to remit than to draw	*il est plus avantageux de remettre que de fournir*
seduc-	it is more advantageous to draw than to remit	*il est plus avantageux de fournir que de remettre*
spand-	open a credit for drafts	*ouvrez un crédit pour traites*
sparg-	opened a credit for drafts	*ouvert un crédit pour traites*
stern-	open a credit for documentary drafts	*ouvrez „ „ „ „ documentaires*
strep-	opened „ „ „ „ „	*ouvert „ „ „ „ „*
subig-	against Letter of Credit for drafts	*contre Lettre de Crédit pour traites*
sublu-	„ „ „ „ „ documentary drafts	*contre Lettre de Crédit pour traites documentaires*
subur-	~~~~~~	~~~~~~

Table 12 Finance.

54 roots ×

pay *payez*	paid *payé*	remit *remettez*	remitted *remis*	draw *fournissez*	received *reçu*	
occup-	profl-	reser-	solid-	suffl-	ulcer-	1
offoc-	pumic-	rigor-	somni-	sugil-	umect-	2
onust-	quadr-	rutil-	sopor-	supin-	ustil-	3
ordin-	recit-	sagin-	spect-	temer-	usurp-	4
perfl-	recus-	salut-	spiss-	tempt-	vapor-	5
plant-	regel-	sauci-	spoli-	terti-	veget-	6
plasm-	relax-	separ-	spurc-	toler-	venen-	7
popul-	relig-	seren-	stagn-	tract	vigil-	8
praed-	resan-	sevoc-	stupr-	tumul-	vocit-	9

27 terminations = 1539 ciphers (Latin). **Table 11.**

		~~	on London *sur Londres*	on / *sur* New York	on / *sur* Paris	on / *sur* Berlin
~~	~~~~	-am	-emini	-entis	-eris	-ito
8 days's.	8 jours de vue	-amus	-endae	—	—	—
30 „ „	30 „ „ „	-ant	—	-entur	-et	—
60 „ „	60 „ „ „	-atis	-endi	-erem	—	—
3 months's.	3 mois „ „	-atur	-endos	-erent	-imur	-itote
4 „ „	4 „ „ „	-ebant	-endum	—	—	—
6 „ „	6 „ „ „	-ebare	-ens	—	—	—
at sight	à vue	-ebas	-entem	-erere	-is	-unto

Examples of Table 11:

impetatis = do not draw, have remitted 60 d. s. — *ne fournissez pas, avons (ai) remis 60 j. v.*
incolendi = shall remit 60 d. s. on London. — *remettrons (remettrai) 60 j. v. s. Londres.*
scriberent = balance 3 m. s. on New York. — *solde 3 m. de vue sur N. Y.*

Examples of Table 12:

rutilatos = remit 375 dollars — *remettez 375 dollars*
praedarere = pay 9500 francs — *payez 9500 francs*
ulcerandos = received 1750 £ — *reçu 1750 £*

37 terminations = 1998 ciphers (Latin) **Table 12.**

	£	francs	dollars	Mks
~~	-abam	—	-ari	—
0	-abant	-andum	-arier	-avero
5	-abare	-ans	-at	-avit
00	-abis	-antem	-atae	-emus
25	-abo	-antis	-atior	-ent
50	-amini	-anto	-atius	-es
75	-amur	-antur	-atos	-etis
000	-andae	-arent	-atote	-etur
500	-andi	-arere	-atum	-or
750	-andos	-ares	-aturi	—

Table 13. Movements *(mouvements)*. 20 roots ×

bend-	discount	*l'escompte*
biad-	exchange	*le change*
biff-	as exchange	*vu que le change*
boll-	freights	*les frêts*
bord-	as freights	*vu que les frêts*
bozz-	steamer freights	*les frêts par vapeur*
brav-	**market**	***marché***
burl-	**prices**	***les prix***
buss-	„ of all qualities	*„ „ de toutes les qualités*
butt-	„ „ current qualities	*„ „ de qualités courantes*
calm-	„ „ superior „	*„ „ de qualités supérieures*
cans-	„ „ raw material	*„ „ de la matière première*
capp-	as prices	*vu que les prix*
catt-	if „	*si les prix*
cazz-	wire whether prices	*télégraphiez si les prix*
cedr-	price	*le prix*
ceff-	as price	*vu que le prix*
cenn-	if „	*si le prix*
chet-	quotations	*les cotes*
cifr-	barsilver	*argent en barres*

Examples:

calmandole = prices of super. qualities continue rising gradually.
les prix de qual. supér. continuent à hausser p. à p.

biadavamo = exchange continues downward tendency.
le change continue en tendance à la baisse.

burlolla = prices are fluctuating considerably.
les prix sont sujet à fluctuations considérables.

Table 13.

		~~	gradually *peu à peu*	rapidly *rapidement*
		←	←	←
are (is) **rising**	***hausse(nt)***	-acelo	-andoci	-aronle
continue(s) rising	*continue(nt) à hausser*	-acene	-andole	-arono
will probably rise	*hausser/ont(a) probablement* .	-aile	-ansene	-aronsi
„ „ shortly rise .	„ „ „ *sous peu*	-aivi	-ansi	-asse
„ „ continue rising	*continuer/ont(a) probablement à hausser*	-ali	-ante	-assero
are (is) **falling**	***baisse(nt)***	-ammo	-antesi	-assimo
continue(s) falling	*continue(nt) à baisser*	-ammolo	-arci	-astene
will probably fall	*baisser/ont(a) probablement* .	-ammone	-are	-astevi
„ „ shortly fall. .	„ „ „ *sous peu*	-ammovi	-arla	-asti
„ „ continue falling	*continuer/ont(a) probablement à baisser*	-ando	-armene	-astici

continue(s) *continue(nt)*	are (is) *sont (est)*		
←	←		
-astila	-erebbe	high	*élevé/e/s*
-atale	-erei	very high	*très élevé/e/s*
-atasi	-eremo	too „	*trop* „
-atela	-eresti	higher	*plus* „
-atemi	-erete	slightly higher	*légèrement plus élevé/e/s*
-atili	-iamla	much „	*beaucoup* „ „
-ato	-iamo	low	*bas/ses*
-atolo	-iamoci	very low	*très bas/ses*
-atone	-iamolo	too „	*trop* „
-atovi	-iamone	lower	*plus* „
-ava	-iate	slightly lower	*légèrement plus bas/ses*
-avaci	-icela	much „	*beaucoup* „ „
-avala	-ila	upward tendency	*en tendance à la hausse*
-avamo	-imelo	downward „	„ „ „ „ *baisse*
-avanlo	-imene	firm	*ferme(s)*
-avate	-imi	firmer	*plus ferme(s)*
-aveli	-ino	very firm	*très* „
-avilo	-oceli	weak	*faible(s)*
-avimi	-ole	very weak	*très faible(s)*
-avine	-olla	fluctuating considerably	*sujet à fluctuations considérable*
-avole	-omele	unaltered	*sans changement*
-avoti	-ossi	nominal	*nominal(es) (nominaux)*
-era	-ovene	normal	*normal(es) (normaux)*
-eranno	-ovi	~~~~~	~~~~~

mov. & stat. 13

Table 14. Movements (*mouvements*). 23 roots ×

→		
inaqu–	the quantity afloat	*le flottant*
inaur–	arrivals	*les arrivages*
incar–	consumption	*la consommation*
incoj–	local consumption	*„ „ locale*
indor–	**demand**	*„* ***demande***
infil–	„ for all qualities	*„ „ pour toutes les qualités*
infun–	„ „ current qualities	*„ „ qualités courantes*
ingel–	„ „ superior „	*„ „ qualités supérieures*
innav–	„ „ freights	*„ „ „ frêts*
innom–	as demand	*vu que la demande*
inond–	local „	*demande locale*
inram–	offerings	*les offres*
insal–	production	*la production*
inset–	receipts in the interior	*les recettes à l'intérieur*
insof–	shipments	*„ embarquements*
inter–	as shipments	*vu que les embarquements*
intuf–	if „	*si les embarquements*
invas–	stocks	*les stocks (existences)*
invol–	our (my) stock	*notre (mon) stock*
irret–	supplies	*les approvisionnements*
irror–	available tonnage	*le tonnage disponible*
iscus–	„ steamer tonnage	*„ „ „ par vapeur*
lacer–		

Examples:

indoravate = { demand will probably shortly decrease rapidly.
la demande diminuera probablement sous peu rapidement.

58 **terminations** = 1334 ciphers (Italian). **Table 14.**

		~~	gradually *peu à peu*	rapidly *rapidement*
		←	←	←
are (is) **increasing**........	***augmente(nt)***...........	-acelo	-are	-atili
continue(s) increasing.....	*continue(nt) à augmenter*	-acene	-arla	-ato
will probably increase	*augmenter/ont(a) probablement*	-aile	-arono	-atolo
„ „ shortly increase	„ „ *sous peu*	-aivi	-asele	-atone
will probably continue increasing......	*continuer/ont(a) probablement à augmenter*	-ali	-asse	-atovi
are (is) **decreasing**	***diminue(nt)***.............	-ammo	-asti	-ava
continue(s) decreasing.....	*continue(nt) à diminuer*	-ando	-atale	-avaci
will probably decrease	*diminuer/ont(a) probablement*	-ansi	-atasi	-avamo
„ „ shortly decrease	„ „ *sous peu*	-ante	-atela	-avate
„ „ continue decreasing.......	*continuer/ont(a) probablement à diminuer*	-arci	-atemi	-aveli

continue(s) *continue(nt)*	are (is) *sont (est)*		
←	←		
-avilo	-imelo	enormous	*énorme/s*
-avimi	-imene	considerable	*considérable(s)*
-avine	-imi	larger	*plus grand/e/s*
-avole	-ino	small	*petit/e/s*
-avoti	-iteli	smaller	*plus petit/e/s*
-era	-oceli	very small	*très „*
-erei	-ole	moderate	*modéré/e/s*
-eremo	-omele	insufficient	*insuffisant/e/s*
-erete	-ommi	exhausted	*épuisé/e/s*
-iamla	-ossi	nil	*nul/le/s*
-iamo	-otelo	almost nil	*à peu près nul/le/s*
-iate	-otti	normal	*normal/e/s (normaux)*
-icela	-ovene	unaltered	*sans changement*
-ila	-ovi	~~~~~	~~~~~

Cable 15 Statistics. 9 roots ×

→		
seb–	**crop** is	*la **récolte** est*
sirp–	„ will probably be	*„ „ sera probablement*
summ–	„ being	*„ „ étant*
supp–	according to reports the crop is	*d'après des rapports la récolte est*
tetr–	„ „ „ „ „ will probably be	*„ „ „ „ „ sera probablement*
trus–	there are reports that the crop is	*il y a des rapports que la récolte est*
urin–	there are reports that the crop will probably be	*„ „ „ „ „ „ „ „ sera probablement*
vetu–	owing to reports of crop being	*par suite de rapports de récolte*
visc–	wire whether you expect that the crop will be	*télégraphiez si vous vous attendez à ce que la récolte sera*

Examples:

suppabamus = accord. to rep. the crop is cons. dam. by cyclone — *d'apr. d. rapp. l. réc. est cnd. cons. p. cyclone*
sirpes = crop will prob. be an early one. — *la réc. sera prob. précoce.*

Cable 16 Statistics. 34 roots ×

		~~	as compared with the average *en comparaison de la moyenne*	as compared with the preceding crop *en comparaison de la récolte précédente*
		→	→	←
the crop	*la **récolte** ....*	lastr–	logor–	merit–
„ „ as to quantity ..	*„ „ quant à quantité*	latin–	lucid–	misur–
„ „ „ „ quality ...	*„ „ quant à qualité.*	lavor–	macin–	modol–
„ „ „ „ quantity & quality ...	*„ „ quant à quantité et qualité .....*	letam–	malus–	monet–
the crop surplus available for export.....	*le surplus de la récolte disponible pour l'exportation*	levit–	menom–	mostr–

the production *la production*	quality *qualité*	shipments *embarquements*		
→	→	→	~~~~~~	~~~~~~
natur–	oscur–	peror–		
niell–	ospit–	piomb–	as compared with last year	*en comparaison de l'année passée*
nomin–	ossid–	popol–	as compared with the average of the season	*en comparaison de la moyenne de la campagne*
nover–	ottur–	pranz–		
olezz–	ovatt–	prest–	of the season as compared with preceding season	*de la campagne en comparaison de la campagne précédente*
—	parod–	—	as compared with sample	*en comparaison de l'échantillon*
onest–	paves–	prezz–	this month as compared with last month	*mois courant en comparaison du mois passé*

Examples:

lavorolla = crop as to qual. will prob. be unsatisf. — *la récolte quant à l. qual. sera prob. non-satisfaisante*
ossidavate = qual. as comp. w. the average will pr. be 15% superior — *la qual. en comp. de la moyenne sera prob. de 15% supérieure*

39 **terminations** = 351 ciphers (Latin). **Table 15.**

considerably damaged *endommagée considérablement*	slightly damaged *endommagée légèrement*	damaged *endommagée*		
←	←	←		
-abam	-amur	-anto	by	*par*
-abamus	-andam	-antur	by cyclone	*par cyclone*
-abant	-andi	-aremus	„ frost	*„ gelée*
-abare	-andior	-arent	„ excessive heat	*„ chaleur excessive*
-abaris	-andius	-arere	by excessive heat and drought	*par chaleur excessive et sécheresse*
-abatur	-andos	-ares	by insects	*par insectes*
-abimur	-ans	-aretis	„ locusts	*„ locustes*
-abis	-antem	-aretur	„ rains	*„ pluies*
-abitis	-antis	-ari	„ excessive rains	*„ pluies excessives*
-abo	-antium	-arier	„ want of rain	*„ manque de pluie*
-amini	—	-at	„ worms	*„ vers*

←		
-atote	abundant	*abondante*
-emus	very abundant	*très abondante*
-ent	a short one	*déficitaire*
-es	an early one	*précoce*
-etis	a late one	*en retard*
-etur	exhausted	*épuisée*
-or	totally destroyed	*totalement anéantie*

53 **terminations** 1802 ciphers (Italian). **Table 16.**

is (are) *est (sont)*	will probably be *sera (seront) probablement*			
←	←			
-acene	-avamo	by / *de* 10 %	superior *supérieur/e/s*	
-aile	-avate	„ 15 „		
-aivi	-avilo	„ 20 „		
-ali	-avimi	„ 25 „		
-ammo	-avine	„ 30 „		
-ando	-avole	„ 40 „		
-ansi	-avoti	„ 50 „		
-ante	-era	„ 10 „	inferior *inférieur/e/s*	
-arci	-erei	„ 15 „		
-are	-eremo	„ 20 „		
-arla	-erete	„ 25 „		
-arono	-iamla	„ 30 „		
-asse	-iamo	„ 40 „		
-asti	-iate	„ 50 „		
-atale	-ila	the same		*le/s(la) même(s)*
-atasi	-imene	about the same		*à peu près le/s (la) même(s)*
-atela	-imi	satisfactory		*satisfaisant/e/s*
-atemi	-ino	more satisfactory		*plus satisfaisant/e(s)*
-atili	-oceli	less „		*moins „*
-ato	-ole	very „		*très „*
-atolo	-olla	unsatisfactory		*non-satisfaisant/e/s*
-atone	-ossi	very unsatisfactory		*très insatisfaisant/e/s*
-atovi	-otelo	insufficient		*insuffisant/e/s*
—	-otti	same as previously estimated		*même(s) qu'estimé/e/s précédem- [ment*
-ava	-ovela	more satisfactory than expected		*plus satisfaisant/e/s qu'anticipé*
-avaci	-ovene	less satisfactory than expected		*moins satisfaisant/e/s qu'anticipé*
-avala	-ovi			

Table 17 Operations. 16 roots ×

aguz-	we (I) accept	*nous acceptons (j'accepte)*
ahog-	buy	*achetez*
apac-	bought	*acheté*
busc-	we (I) can conclude	*nous pouvons (je peux) conclure [clure*
caz-	„ „ „ probably conclude	*nous (je) pourr/ons(ai) probablement con-*
cerc-	maximum	*maximum*
cruz-	minimum	*minimum*
drag-	try to make us (me) firm offer for	*tâchez de nous (me) faire offre ferme pour*
driz-	we (I) make you a firm offer for immediate telegraphic reply	*nous v. faisons (je v. fais) offre ferme pour réponse télégraph. immédiate*
falc-	wire best possible price for	*télégraphiez meilleur prix possible pour*
fisg-	„ „ „ „ and shortest possible delay of delivery for	*télégraphiez meilleur prix possible et délai le plus bref de livraison pour*
freg-	sell	*vendez*
hinc-	sold	*vendu*
lanz-	ship (send)	*expédiez*
pesc-	shipped (sent)	*expédié*
pleg-	we (I) can probably ship (send)	*nous (je) pourr/ons(ai) probablem. expédier*

Table 18 Operations. 16 roots ×

sapor-	we (I) accept	*nous acceptons (j'accepte)*
sband-	buy	*achetez*
scarn-	bought	*acheté*
scerp-	we (I) can conclude	*nous pouvons (je peux) conclure [clure*
scult-	„ „ „ probably conclude	*nous (je) pourr/ons(ai) probablement con-*
secur-	maximum	*maximum*
semin-	minimum	*minimum*
spalm-	try to make us (me) firm offer for	*tâchez de nous (me) faire offre ferme pour*
sparl-	we (I) make you a firm offer for immediate telegraphic reply	*nous v. faisons (je v. fais) offre ferme pour réponse télégraph. immédiate*
spazz-	wire best possible price for	*télégraphiez meilleur prix possible pour*
spegn-	„ „ „ „ and shortest possible delay of delivery for	*télégraphiez meilleur prix possible et délai le plus bref de livraison pour*
spens-	sell	*vendez*
stamp-	sold	*vendu*
stend-	ship (send)	*expédiez*
still-	shipped (sent)	*expédié*
stirp-	we (I) can probably ship (send)	*nous (je) pourr/ons(ai) probablement expédier*

Table 19 Operations. 16 roots ×

stizz-	we (I) accept	*nous acceptons (j'accepte)*
storn-	buy	*achetez*
strem-	bought	*acheté*
sturb-	we (I) can conclude	*nous pouvons (je peux) conclure [clure*
tacit-	„ „ „ probably conclude	*nous (je) pourr/ons(ai) probablement con-*
tempr-	maximum	*maximum*
tesor-	minimum	*minimum*
titol-	try to make us (me) firm offer for	*tâchez de nous (me) faire offre ferme pour*
tutel-	we (I) make you a firm offer for immediate telegraphic reply	*nous v. faisons (je v. fais) offre ferme pour réponse télépraph. immédiate*
umett-	wire best possible price for	*télégraphiez meilleur prix possible pour*
uncin-	„ „ „ „ and shortest possible delay of delivery for	*télégraphiez meilleur prix possible et délai le plus bref de livraison pour*
unizz-	sell	*vendez*
valor-	sold	*vendu*
valut-	ship (send)	*expédiez*
velen-	shipped (sent)	*expédié*
vidim-	we (I) can probably ship (send)	*nous (je) pourr/ons(ai) probablement expédier*

56 **terminations** = 896 ciphers (Spanish) **Table 17.**

	~~	00	000 yards	00 meters *00 mètres*
1	-abais-	-adoles	-arame	-arnos
2	-abalas	-adonos	-aramos	-aron
3	-abalo	-ados	-aranlo	-aronla
4	-aban	-adote	-arante	-aronse
5	-abanla	-ais	-aras	-asemos
10	-abanme	-ales	-areis	-asen
15	-abanos	-andolo	-arelas	-aste
20	-abase	-andome	-arelo	-asteis
25	-ada	-anla	-arete	-astela
50	-adale	-anlos	-aria	-asteme
100	-adlas	-anse	-ariais	-olas
150	-adlo	-ar	-arialo	-olo
200	-adme	-arala	-ariase	-ome
250	-adola	-arales	-arle	-onos

Examples:
hincaras = { sold 5000 yds. — *rendu 5000 yds.*
lanzarala = { ship (send) 20000 — *expédiez 20000.*

60 **terminations** = 960 ciphers (Italian) **Table 18.**

	bags *sacs*	bales *balles*	cases *caisses*	barrels, casks, kegs *barils, fûts etc.*
50	-acelo	-asti	-aveli	-imelo
100	-acene	-atale	-avilo	-imene
125	-aile	-atasi	-avimi	-imi
150	-aivi	-atela	-avine	-ino
200	-ali	-atemi	-avole	-iteli
250	-ammo	-atili	-avoti	-oceli
300	-ando	-ato	-era	-ole
500	-ansi	-atolo	-erei	-olla
1000	-ante	-atone	-eremo	-omele
1250	-arci	-atovi	-erete	-ossi
1500	-are	-ava	-iamla	-otelo
2000	-arla	-avaci	-iamo	-otti
2500	-arono	-avala	-iate	-ovela
3000	-asele	-avamo	-icela	-ovene
5000	-asse	-avate	-ila	-ovi

Examples:
sbandacene = { buy 100 bags. — *achetez 100 sacs.*
stirpatone = { we (I) can prob. ship (send) 1000 bales. — *nous (je) pourr. prob. exp. 1000 balles.*

60 **terminations** = 960 ciphers (Italian) **Table 19.**

oper^ns 17

	tons of 20 cwt *tonnes de 20 cwt*	cwt	lbs	Kilos
10	-acelo	-asti	-aveli	-imelo
25	-acene	-atale	-avilo	-imene
50	-aile	-atasi	-avimi	-imi
100	-aivi	-atela	-avine	-ino
150	-ali	-atemi	-avole	-iteli
200	-ammo	-atili	-avoti	-oceli
250	-ando	-ato	-era	-ole
300	-ansi	-atolo	-erei	-olla
500	-ante	-atone	-eremo	-omele
1000	-arci	-atovi	-erete	-ossi
1250	-are	-ava	-iamla	-otelo
1500	-arla	-avaci	-iamo	-otti
2000	-arono	-avala	-iate	-ovela
2500	-asele	-avamo	-icela	-ovene
3000	-asse	-avate	-ila	-ovi

Examples:
valutammo = { ship (send) 200 tons of 20 cwt. — *expédiez 200 tonnes de 20 cwt.*

Table 20 Operations. 54 roots ×

Buying, Selling, Offering &c. *Achats, Ventes, Offres etc.*

		buy *acheter*	sell *vendre*
we (I) recommend you to	*nous vous recommandons (je vous recommande) d(e)*	abaj-	ajor-
„ „ „ „ at once to	*nous vous recommandons (je vous recommande) d'immédiatement*	abon-	alej-
it seems advisable to	*il semble à propos d(e)*	abur-	alif-
„ „ „ not to	*il semble à propos de ne pas*	aced-	alod-
wire whether we (I) may	*télégraphiez si nous pouvons (je puis)*	acit-	alun-
„ „ you recommend us (me) to	*„ „ vous nous (me) recommandez d(e)*	acos-	amas-
do not	*abstenez-vous d(e)*	actu-	amer-
continue to	*continuez à*	adem-	amim-
do not continue to	*ne continuez pas à*	adob-	anex-
we (I) can	*nous pouvons (je peux)*	aduj-	anud-
„ „ „ probably	*„ (je) pourr/ons(ai) probablement*	afin-	ape-
wire whether you can	*télégraphiez si vous pouvez*	afof-	apip-
we (I) cannot	*nous ne pouvons (je ne puis)*	afum-	apoy-
„ „ can probably not	*„ „ pourrons (je ne pourrai) probablement pas*	ahij-	apro-
„ „ intend to	*nous avons (j'ai) l'intention d(e)*	ahus-	arcu-
„ „ do not intend to	*nous n'avons (je n'ai) pas l'intention d(e)*	ajen-	arel-

asil-	we are (I am) buying	*nous achetons (j'achète)*
atet-	buy	*achetez*
atop-	sell	*vendez*
atuf-	we (I) have bought	*nous avons (j'ai) acheté*
aupr-	„ „ „ sold	*„ „ „ vendu*
avan-	provided you buy	*pourvu que vous achetiez*
aviv-	„ „ sell	*„ „ „ vendiez*
azot-	unless you buy	*à moins que vous n'achetiez*
azul-	„ „ sell	*„ „ „ „ ne vendiez*
bald-	there are **buyers**	*il y a* ***acheteurs***
bebd-	„ „ sellers	*„ „ „ vendeurs*
bef-	„ „ no buyers	*„ n'y a pas d'acheteurs*
beix-	„ „ „ sellers	*„ „ „ „ de vendeurs*
birl-	„ „ many buyers	*„ y a beaucoup d'acheteurs*
bizm-	„ „ „ sellers	*„ „ „ „ de vendeurs*
boch-	**competitors** have bought	*la* ***concurrence*** *a acheté*
borr-	„ „ sold	*„ „ „ vendu*
bosl-	„ are offering	*„ „ offre*
boy-	wire **limit**	*télégraphiez* ***limite***
bras-	wire limits	*„ limites*
brot-	endeavour to make us (me) **firm offer**	*tâchez de nous (me) faire* ***offre*** *ferme*
brum-	we (I) make you firm offer for immediate telegraphic reply	*nous vous faisons (je v. fais) offre ferme pour réponse télégraphique immédiate*

Examples:

aburarante = it seems advisable to buy at these prices / *il semble à pr. d'ach. à ces prix*

befabalo = there are no buyers shipment next month / *il n'y a pas d'acheteurs expédition mois proch.*

amasados = wire wh. you recommend us (me) to sell / *télégr. si vous nous (me) recomm. de vendre*

65 **terminations** = 3510 ciphers (Spanish) **Table 20.**

shipment *expédition*	**delivery** *livraison*		
←	←		
-abais	-abase	immediately	*immédiate*
-abalas	-ada	present month	*mois courant*
-abalo	-adale	next „	*„ prochain*
-aban	-adlas	present and next month	*„ courant et prochain*
-abanla	-adlo	next 2 months	*les 2 mois prochains*
—	-adme	15th of present month	*le 15 du mois courant*
—	-adola	15th „ next „	*„ 15 „ „ prochain*
-abanme	-adoles	end „ present „	*fin mois courant*
-abanos	-adonos	„ „ next „	*„ „ prochain*

←		
-ados	~~~~~~~~	~~~~~~~~
-adote	to arrive	*à arriver*
-ais	at auction	*aux enchères*
-ales	to cover	*en couverture*
-andolo	new crop	*nouvelle récolte*
-andome	old „	*ancienne „*
-anla	on the spot	*sur place*
-anlos	at this **price**	*à ce **prix***
-anse	„ the price limited	*au prix limité*
-ar	„ „ „ offered	*„ „ offert*
-arala	„ previous price	*„ „ antérieur*
-arales	„ the price mentioned	*„ „ mentionné*
-arame	„ „ same price (as)	*„ même prix (que)*
-aramos	„ a higher price	*à un prix plus élevé*
-aranlo	„ „ lower „	*„ „ „ „ bas*
-arante	„ these **prices**	*„ ces prix*
-aras	„ prices limited	*aux prix limités*
-areis	„ „ offered	*„ „ offerts*
-arelas	„ „ mentioned	*„ „ mentionnés*
-arelo	„ present prices	*„ „ actuels*
-arete	„ same prices (as)	*„ mêmes prix (que)*
-aria	„ higher prices	*à prix plus élevés*
-ariais	„ lower „	*„ „ „ bas*
-arialo	A I **quality**	***qualité** A I*
-ariase	superior quality	*„ supérieure*
-arle	current „	*„ courante*
-arnos	inferior „	*„ inférieure*
-aron	quality = to sample	*„ = à l'échantillon*
-aronla	„ about = to sample	*„ à peu près = à l'échantillon*
-aronse	„ mentioned	*„ mentionnée*
-asemos	same quality (as)	*même qualité (que) [la dernière fois*
-asen	„ „ and same terms as last time	*„ „ et mêmes conditions que*
-aste	important **quantity**	***quantité** importante*
-asteis	small „	*petite quantité*
-astela	larger „	*quantité plus grande*
-asteme	smaller „	*„ „ petite*
-ela	double „	*double quantité*
-elos	half „	*la moitié*
-emonos	the remainder	*le reste*
-en	total quantity	*quantité totale*
-enlas	quantity mentioned [last time	*„ mentionnée [que dernière fois*
-enlo	same quantity and same terms as	*même quantité et mêmes conditions*
-enme	on same **terms** (as)	*aux mêmes **conditions** (que)*
-ennos	„ these „	*à ces conditions*
-ese	„ usual „	*aux conditions habituelles*
-olas	„ the terms mentioned	*„ „ mentionnées*
-olo	fob	*franco bord*
-ome	c. & f.	*coût et frêt*
-onos	cif	*coût, frêt et assurance*

Table 21.

16 roots ×

Present and Future Position of Prices.
Situation Actuelle et Future des Prix.

→		
buh-	**buy** as	***achetez*** *vu que*
bull-	it seems advisable to buy at once as	*il semble à propos d'acheter immédiatement vu que*
burx-	buy further as	*achetez davantage vu que*
call-	it seems advisable to cover as	*il semble à propos de couvrir vu que*
cand-	there are buyers as	*il y a acheteurs vu que*
capt-	„ „ many buyers as	*„ „ „ beaucoup d'acheteurs vu que*
caus-	try to make firm offer as	*tâchez de faire offre ferme vu que*
chaf-	we (I) make you the following firm offer for immediate telegr. reply as	*nous v. faisons (je v. fais) l'offre ferme suivante pour réponse télégraphique immédiate vu que*
chom-	wire limits for buying as	*télégraphiez limites pour acheter vu que*
chup-	do **not sell** as	*ne* ***vendez pas*** *vu que*
cicl-	it seems advisable not to sell at present as	*il semble à propos de ne pas vendre en ce moment vu que*
cint-	we (I) do not intend to sell at present as	*nous n'avons (je n'ai) pas l'intention de vendre en ce moment vu que*
clav-	„ „ have not sold as	*nous n'avons (je n'ai) pas vendu vu que*
clis-	there are no sellers as	*il n'y a pas de vendeurs vu que*
cobr-	sellers are reserved as	*les vendeurs sont réservés vu que*
coed-	~~~~~~	~~~~~~

Examples:

cobraronse = sellers are reserved as prices are very low.
les vend. s. réservés vu que les prix sont très bas.

bullada = it seems adv. to buy at once as prices will prob. rise owing to decrease in shipts. — *il semble à propos d'ach. imméd. vu que les prix hausseront prob. par suite de dimin. d'emb.*

65 **terminations** = 1040 ciphers (Spanish) **Table 21.**

prices will probably **rise** *les prix* ***hausseront*** *probablement*					
owing to: *par suite de:*	**shortly** owing to: ***sous peu*** *par suite de:*	**considerably** owing to: ***considé-rablement*** *par suite de:*	in view of **probability** of: *en la* ***proba-bilité*** *de:*		
←	←	←	←		
-abais	-adlo	-anla	-areis	unfavorable crop accounts	*rapports défavorables de récolte*
-abalas	-adme	-anlos	-arelas	less favorable crop accounts	*rapports moins favorables de récolte*
-abalo	-adola	-anse	-arelo	crop damage	*dommage de récolte*
-aban	-adoles	-ar	-arete	increase in demand	*augmentation de demande*
-abanla	-adonos	-arala	-aria	strong demand	*forte demande*
-abanme	-ados	-arales	-ariais	decrease in production	*diminution de production*
-abanos	-adote	-arame	-arialo	considerable purchases	*achats considérables*
-abase	-ais	-aramos	-ariase	decrease in receipts in the interior	*diminution de recettes à l'intérieur*
-ada	-ales	-aranlo	-arle	decrease in shipments	*diminution d'embarquements*
-adale	-andolo	-arante	-arnos	decrease in stocks	*diminution de stocks*
-adlas	-andome	-aras	-aron	small stocks	*petits stocks*

prices are: *les prix sont:*	prices are reported: *les prix sont signalés:*	prices will probably be: *prix seront probablement*		
←	←	←		
-aronla	-asteme	-enme	low	*bas*
-aronse	-ela	-ennos	very low	*très bas*
-asemos	-elos	-ese	exceedingly low	*excessivement bas*
-asen	-emonos	-olas	much lower	*beaucoup plus bas*
-aste	-en	-olo	rising	*en hausse*
-asteis	-enlas	-ome	„ gradually	„ „ *graduelle*
-astela	-enlo	-onos	„ rapidly	„ „ *rapide*

Cable 22.

16 roots X

Present and Future Position of Prices.
Situation Actuelle et Future des Prix.

colm–	**sell** as	***vendez*** *vu que*
comb–	we (I) recommend you to sell at once as	*nous (je) vous recommend/ons(e) de vendre immédiatement vu que*
corl–	it seems advisable to sell at once as	*il semble à propos de vendre immédiatement vu que*
coyt–	sell further as	*vendez davantage vu que*
crib–	there are sellers as	*il y a vendeurs vu que*
cuaj–	„ „ many sellers as	*„ „ „ beaucoup de vendeurs vu que*
cub–	try to make firm offer as	*tâchez de faire offre ferme vu que*
cuid–	we (I) make you the following firm offer for immediate telegraphic reply as	*nous v. faisons (je v. fais) l'offre ferme suivante pour réponse télégraph. immédiate vu que*
curs–	wire limits for selling as	*télégraphiez limites pour vendre vu que*
dabi–	do **not buy** as	*n'**achetez pas** vu que*
dan–	it seems advisable not to buy at present as	*il semble à propos de ne pas acheter en ce moment vu que*
dej–	we (I) do not intend to buy at present as	*nous n'avons (je n'ai) pas l'intention d'acheter en ce moment vu que*
dezm–	we (I) have not bought as	*nous n'avons (je n'ai) pas acheté vu que*
dubd–	there are no buyers as	*il n'y a pas d'acheteurs vu que*
dup–	buyers are reserved as	*les acheteurs sont réservés vu que*
edr–	~~~~~~	~~~~~~

Examples:

dubdarete = there are no buyers as prices will prob. decline in view of prob. of decr. in demand. — *il n'y a pas d'achet. vu que les prix baiss. prob. vu la probabilité de diminution de demande.*

edrarame = prices will prob. decl. cons. owing to overproduction. *les prix baiss. prob. cons. par suite de surprod.*

prices will probably **decline** / *les prix* ***baisseront*** *probablement*					
owing to: *par suite de:*	shortly owing to: ***sous peu*** *par suite de:*	considerably owing to: ***considérablement*** *par suite de:*	in view of probability of: *vu la* ***probabilité*** *de:*		
←	←	←	←		
-abais	-adlo	-anla	-areis	favorable crop accounts	*rapports favorables de récolte*
-abalas	-adme	-anlos	-arelas	more favorable crop accounts	*„ plus favorables de récolte*
-abalo	-adola	-anse	-arelo	insignificant demand	*demande insignifiante*
-aban	-adoles	-ar	-arete	decrease in demand	*diminution de demande*
-abanla	-adonos	-arala	-aria	considerable offerings	*offres considérables*
-abanme	-ados	-arales	-ariais	„ production	*production considérable*
-abanos	-adote	-arame	-arialo	overproduction	*surproduction*
-abase	-ais	-aramos	-ariase	considerable realizations	*réalisations considérables*
-ada	-ales	-aranlo	-arle	increase in receipts in the interior	*augmentation de recettes à l'intérieur*
-adale	-andolo	-arante	-arnos	increase in shipments	*augmentation d'embarquements*
-adlas	-andome	-aras	-aron	„ „ stocks	*augmentation de stocks*

prices are: *les prix sont:*	prices are reported: *les prix sont signalés:*	prices will probably be: *les prix seront probablement:*		
←	←	←		
-aronla	-asteme	-enme	high	*élevés*
-aronse	-ela	-ennos	very high	*très élevés*
-asemos	-elos	-ese	exceedingly high	*excessivement élevés*
-asen	-emonos	-olas	much higher	*beaucoup plus élevés*
-aste	-en	-olo	declining	*en baisse*
-asteis	-enlas	-ome	„ gradually	*„ „ graduelle*
-astela	-enlo	-onos	„ rapidly	*„ „ rapide*

Table 23 +. 23 roots ×

→		
cign–	at	*à*
cim–	there are **buyers** at	*il y a* ***acheteurs*** *à*
clam–	„ „ sellers „	*„ „ „ vendeurs à*
colp–	competitors have bought	*la concurrence a acheté*
corn–	„ „ sold	*„ „ „ vendu*
crep–	„ are offering	*„ „ offre*
cuit–	we (I) have **concluded** at	*nous avons (j'ai)* ***conclu*** *à*
cull–	we (I) can conclude at	*nous pourons (je peux) conclure à*
curr–	„ „ „ probably conclude at	*„ pourrons (je pourrai) probablement conclure à*
dann–	with a firm offer in hand we (I) can probably conclude **at**	*avec une offre ferme en mains nous pourrons (je pourrai) probablement conclure à*
degn–	try to conclude at	*tâchez de conclure à*
ditt–	impossible to do any business unless at	*impossible de faire des affaires sinon à*
esil–	we (I) fix our (my) **limit**	*nous fixons (je fixe) notre (ma)* **limite**
faon–	„ „ „ „ „ limits	*„ „ „ „ nos (mes) limites*
ferm–	**market** is	***marché*** *est*
fet–	our **prices** are	*nos* ***prix*** *sont*
fiss–	we (I) accept **offer** at ..... and wait your immediate telegraphic reply	*nous acceptons (j'accepte)* **offre** *à .... et attendons (attends) votre réponse télégraphique immédiate*
fiut–	we (I) accept offer at ..... and shall consider business concluded, unless we (I) hear from you to the contrary by your immediate telegraphic reply.	*nous acceptons (j'accepte)) offre à .... et considérerons (considérerai) l'affaire conclue sauf votre avis contraire par réponse télégraphique immédiate*
frod–	try to renew offer at	*tâchez de renouveler offre à*
fru–	we (I) renew firm offer for immediate telegraphic reply at	*nous renouvelons (je renouvelle) offre ferme pour réponse télégr. immédiate à*
furt–	we (I) make you firm offer for immediate telegraphic reply at	*nous vous faisons (je vous fais) offre ferme pour réponse télégr. immédiate à*
gabb–	against your firm offer we (I) make you counter-offer for immediate telegraphic reply at	*contre v. offre ferme nous vous faisons (je vous fais) contre-offre pour réponse télégraphique immédiate à*
garz–	**option** .... for	***option*** *.... pour*

Examples:

fermandole = market is **3** d. lower (cheaper). — *marché est 3 d. plus bas (m. m.)*

gabbavoti = against your firm offer we (I) make you counter-offer for imm. tel. reply at $1^1/_2$ d. higher (dearer). — *contre v. offre ferme nous v. faisons (je vous fais) contre-offre p. rép. tél. imm. à $1^1/_2$ d. plus élevé (p. ch.)*

83 **terminations** = 1909 ciphers (Italian). **Table 23.**

	lower (cheaper) **plus bas** *(meilleur marché)*	**higher** (dearer) **plus élevé/s** *(plus cher/s)*
	←	←
1/32 d.	-acelo	-avala
1/16 „	-acene	-avamo
3/32 „	-aile	-avanlo
1/8 „	-aivi	-avate
1/4 „	-ali	-aveli
3/8 „	-ammo	-avilo
1/2 „	-ammolo	-avimi
3/4 „	-ammone	-avine
1 „	-ammovi	-avole
1 1/2 „	-ando	-avoti
2 „	-andoci	-era
3 „	-andole	-eranno
4 „	-ansene	-erebbe
4 1/2 „	-ansi	-erei
6 „	-ante	-eremo
1/-	-antesi	-eresti
1/3	-arci	-erete
1/6	-are	-iamla
2/-	-arla	-iamo
0,01	-armene	-iamoci
0,02 1/2	-aronle	-iamolo
0,05	-arono	-iamone
0,10	-aronsi	-iate
0,25	-asse	-icela
0,50	-assero	-ila
0,75	-assimo	-imelo
1,—	-astene	-imene
1/4 %	-astevi	-imi
1/2 „	-asti	-ino
3/4 „	-astici	-oceli
1 „	-astila	-ole
1 1/2 „	-atale	-olla
2 „	-atasi	-omele
2 1/2 „	-atela	-ommi
3 „	-atemi	-onne
3 1/2 „	-atili	-ossi
4 „	-ato	-otelo
5 „	-atolo	-otti
6 „	-atone	—
7 1/2 „	-atovi	-ovela
10 „	-ava	-ovene
~~	-avaci	-ovi

+ 23

Table 24.

63 roots ×

Prices *(amounts)* from **0** to **20** with **32** nds **16** ths and **8** ths

cents	d.	£ stg		cents	d.	£ stg	
abige-	argut-	coact-	~	divex-	exsec-	infor-	11
accit-	assol-	crisp-	1	effet-	facet-	infri-	12
adhal-	atton-	cunct-	2	effoc-	ferit-	inhum-	13
adlev-	audit-	dapin-	3	elect-	fluid-	iniqu-	14
advel-	balte-	dedom-	4	elong-	fodic-	inorn-	15
aemul-	camer-	degul-	5	emanc-	ignav-	inton-	16
aestu-	camps-	delav-	6	emusc-	inalb-	jubil-	17
albic-	capri-	deocc-	7	evers-	incal-	jurgi-	18
allab-	capul-	deopt-	8	exalb-	incib-	litur-	19
amoen-	carin-	desec-	9	exjur-	inerr-	livid-	20
angul-	cernu-	devel-	10				

Table 25.

100 roots ×

decimal prices *(amounts)* from **1** to **100**

giel-	1	merl-	21	popp-	41	sest-	61	tarp-	81
giov-	2	mirr-	22	pred-	42	sevr-	62	tast-	82
giur-	3	mur-	23	pugn-	43	siep-	63	tent-	83
glos-	4	mutu-	24	punt-	44	situ-	64	terz-	84
gol-	5	nasp-	25	quet-	45	smel-	65	tos-	85
gram-	6	nerb-	26	raff-	46	smir-	66	trit-	86
grid-	7	nett-	27	ramp-	47	snod-	67	trov-	87
guad-	8	not-	28	ras-	48	sogn-	68	tuff-	88
inaj-	9	nud-	29	raun-	49	soj-	69	uman-	89
indu-	10	onor-	30	ren-	50	solf-	70	util-	90
inum-	11	orl-	31	riar-	51	sopr-	71	vant-	91
lard-	12	pel-	32	rizz-	52	sors-	72	var-	92
libr-	13	pern-	33	rob-	53	spaj-	73	vess-	93
liev-	14	pian-	34	rum-	54	spin-	74	viet-	94
lod-	15	pill-	35	sacr-	55	spos-	75	voit-	95
lott-	16	pinz-	36	salt-	56	spul-	76	vuot-	96
mamm-	17	piot-	37	scaf-	57	staz-	77	zepp-	97
mand-	18	pip-	38	scip-	58	stiv-	78	zinn-	98
marn-	19	pium-	39	scor-	59	stut-	79	zomb-	99
mass-	20	pond-	40	scud-	60	suon-	80	zon-	100

Table 26.

101 roots ×

other prices *(amounts)* and quantities from **1** to **100**

abit-	1	eman-	22	livr-	42	ponz-	62	sput-	82
adim-	2	equ-	23	lord-	43	pot-	63	stas-	83
adot-	3	ered-	24	lut-	44	prov-	64	stir-	84
adus-	4	esit-	25	magn-	45	quot-	65	stuf-	85
alz-	5	esor-	26	mans-	46	rap-	66	suzz-	86
baj-	6	esum-	27	marz-	47	rub-	67	tass-	87
bard-	7	fid-	28	mert-	48	rutt-	68	tond-	88
beff-	8	forz-	29	mitr-	49	sagr-	69	topp-	89
biul-	9	ganz-	30	mozz-	50	sald-	70	trem-	90
bram-	10	gess-	31	mund-	51	san-	71	trin-	91
bus-	11	gitt-	32	nas-	52	sarp-	72	trot-	92
caut-	12	grif-	33	noj-	53	scav-	73	turb-	93
cepp-	13	guid-	34	odor-	54	scoj-	74	urt-	94
cerr-	14	inam-	35	olf-	55	scus-	75	vent-	95
cib-	15	inor-	36	oltr-	56	segn-	76	vers-	96
coll-	16	isol-	37	ombr-	57	sod-	77	vign-	97
cur-	17	laid-	38	orm-	58	somm-	78	volt-	98
dam-	18	latt-	39	pann-	59	sost-	79	zaff-	99
dett-	19	lem-	40	pens-	60	spar-	80	zamp-	100
dobl-	20	lenz-	41	pest-	61	spes-	81	zel-	~
dos-	21								

28 terminations = 1764 ciphers (Latin) **Table 24.**

-abam	1/32	-andum	3/8	-ari	11/16
-abant	1/16	-ans	13/32	-arier	23/32
-abare	3/32	-antem	7/16	-at	25/32
-abis	1/8	-antis	15/32	-atote	13/16
-abo	5/32	-anto	17/32	-emus	27/32
-amini	3/16	-antur	9/16	-ent	7/8
-amur	7/32	-arent	19/32	-es	29/32
-andae	9/32	-arere	5/8	-etis	15/16
-andi	5/16	-ares	21/32	-etur	31/32
-andos	11/32				

Examples:
balteabis = 4 1/8 d. — inhumabam = £ stg 13 1/32

80 terminations = 8000 ciphers (Italian) **Table 25.**

~~	dollar/s	franc/s	Mark/s *(Marc/s)*	
-acelo	-aronle	-ava	-iamo	,00
-acene	-arono	-avaci	-iamoci	,05
-aile	-aronsi	-avala	-iamolo	,10
-aivi	-asse	-avamo	-iamone	,15
-ali	-assero	-avanlo	-iate	,20
-ammo	-assimo	-avate	-icela	,25
-ammolo	-astene	-aveli	-ila	,30
-ammone	-astevi	-avilo	-imelo	,35
-ammovi	-asti	-avimi	-imene	,40
-ando	-astici	-avine	-imi	,45
-andoci	-astila	-avole	-ino	,50
-andole	-atale	-avoti	-oceli	,55
-ansene	-atasi	-era	-ole	,60
-ansi	-atela	-eranno	-olla	,65
-ante	-atemi	-erebbe	-omele	,70
-antesi	-atili	-erei	-ossi	,75
-arci	-ato	-eremo	-otelo	,80
-are	-atolo	-eresti	-otti	,85
-arla	-atone	-erete	-ovene	,90
-armene	-atovi	-iamla	-ovi	,95

Examples: quetammolo = 45,30 — vanteranno = frs 91,65

65 terminations = 6565 ciphers (Italian) **Table 26.**

-acela	1/4
-ala	1/3
-amelo	1/2
-ammoci	2/3
-ammola	3/4
-andomi	bags *(sacs)*
-andone	bales *(balles)*
-anle	barrels (casks, kegs) *barils, fûts etc.*
-ano	bushels
-antene	cases *(caisses)*
-anti	% (per cent)
-arlo	& 1/2 % (per cent)
-armela	days *(jours)*
-armi	doz. *(douz.)*
-arne	feet
-aronvi	„ square
-arsene	cbft

-assi	gallons
-astele	gross/*es*
-astemi	in.
-astilo	meter *(mè-*
-astine	m² *tres)*
-atala	m³
-atami	c/m
-atane	m/m
-ate	yards

Currency

-ateci	centimes
-atelo	cents
-atile	1/4 cents
-atisi	1/2 „
-atoti	3/4 „
-avale	fl.
-avanci	d. (pence)

-avanla	1/4 d.
-avano	1/2 „
-avasi	3/4 „
-avi	Pfg.
-avici	£
-avila	„ 5 s.
-avolo	„ 10 „
-avomi	„ 15 „
-avone	s. (shill.)
-erai	s. 1 d.
-eranlo	„ 1 1/2 „
-erassi	„ 2 „
-eremmo	„ 3 „
-ereste	„ 4 „
-ero	„ 4 1/2 „
-iamlo	„ 5 „
-iamola	„ 6 „

-iamoti	s. 7 d.
-iamvi	„ 7 1/2 „
-iceli	„ 8 „
-icene	„ 9 „
-ici	„ 10 „
-ilo	„ 10 1/2 „
-imela	„ 11 „

Weights

-occi	tons of 20 cwt
-ocelo	„ „ 2000 lbs
-ollo	cwt
-omene	qr (quarters)
-omi	lb
-one	oz. *(onces)*
-otele	Kilo
-oveli	tons of 1000 Kilo *(tonnes de „ „)*

Examples:
ruberassi = 67/2 — equavasi = 23 3/4 d — turbammoci = 93 2/3

quant. 24

Table 27. 91 roots ×

Quantities *(amounts)* from **100** to **9950**

		chas–	31	guin–	55	palm–	79
		chor–	32	hall–	56	pif–	80
		chum–	33	hart–	57	pun–	81
ases–	10	cort–	34	hen–	58	quel–	82
asin–	11	crin–	35	iris–	59	quit–	83
asob–	12	cuch–	36	jorr–	60	ream–	84
asum–	13	cudi–	37	jur–	61	redr–	85
ativ–	14	dol–	38	lacr–	62	rent–	86
ator–	15	enoj–	39	legr–	63	robl–	87
atus–	16	evol–	40	lep–	64	sach–	88
aven–	17	fad–	41	lidi–	65	sed–	89
avis–	18	fech–	42	list–	66	sign–	90
azem–	19	feri–	43	los–	67	test–	91
bali–	20	feud–	44	merm–	68	tild–	92
beld–	21	fij–	45	mes–	69	tint–	93
bin–	22	foli–	46	mim–	70	trat–	94
bond–	23	frem–	47	mirl–	71	turr–	95
born–	24	frot–	48	moch–	72	vend–	96
brid–	25	fum–	49	moll–	73	vici–	97
brom–	26	fund–	50	moy–	74	vilt–	98
cabs–	27	gan–	51	mud–	75	vir–	99
calv–	28	gast–	52	nubl–	76	zaf–	∼
card–	29	gom–	53	orf–	77		
ceb–	30	grab–	54	pact–	78		

Examples:

bridandote = 255 £ stg — tintamos = 935 frs — signose = 9050

Table 28. 92 roots ×

Quantities *(amounts)* from **10000** to **10 050 000**

		admir–	31	alamp–	55	aquej–	79
		aduan–	32	alarm–	56	ardin–	80
		adver–	33	alaud–	57	argil–	81
abebr–	10	afect–	34	alegr–	58	argur–	82
abism–	11	afelp–	35	alert–	59	arpon–	83
abord–	12	afibl–	36	algam–	60	arrap–	84
abrev–	13	afloj–	37	alind–	61	arrid–	85
abrom–	14	afond–	38	alipi–	62	arroj–	86
abrut–	15	agach–	39	alist–	63	arruf–	87
abult–	16	aglob–	40	almen–	64	asalt–	88
acens–	17	agobi–	41	aloit–	65	ascus–	89
acept–	18	agram–	42	amain–	66	asign–	90
acerr–	19	agris–	43	amant–	67	asoci–	91
achin–	20	agrup–	44	amort–	68	asorb–	92
acich–	21	aguij–	45	anhil–	69	aspas–	93
aclam–	22	ahetr–	46	anill–	70	atabl–	94
acopl–	23	ahirm–	47	anubl–	71	atair–	95
activ–	24	ahorn–	48	apand–	72	atedi–	96
acuci–	25	ahuch–	49	apart–	73	atenu–	97
adarv–	26	ahull–	50	apell–	74	atigr–	98
adecu–	27	ajegu–	51	apest–	75	atild–	99
adeud–	28	ajust–	52	aplan–	76	atipl–	100
adiet–	29	alaci–	53	apret–	77	atoch–	∼
adjur–	30	aladr–	54	apunt–	78		

56 **terminations** = 5096 ciphers (Spanish) **Table 27.**

0	5	00	50	
←	←	←	←	
-aba	-amonos	-aranla	-enles	dollars
-abala	—	-aranme	—	florins
-abales	-amos	-aranos	-enos	frs
—	—	-arate	—	Lire it.
-abame	-andola	-arela	-ense	marks
—	—	-areles	—	milreis
—	—	-areme	—	piaster
-abamos	-andote	-aremos	-ete	£ stg
—	—	-aren	—	rubles
—	—	-ariala	—	rupees
—	—	-ariame	—	taels
—	—	-arian	—	yen
-abanlo	-anlas	-arieis	—	bags *(sacs)*
-abaste	-anlo	-arlos	—	bales *(balles)*
-ad	—	-aronlo	-ola	barrels, casks or kegs *barils, fûts etc.*
-adas	—	-aronte	—	bushels
-adla	-anme	-arse	—	cases *(caisses)*
-adles	—	-aseis	—	gallons
-adnos	-arais	-astelo	-oles	Kilos
-adolas	—	—	—	qr (quarters)
-adolo	-aralas	-elas	-omelos	tons of 20 cwt
-adome	—	-elo	—	yards
-ale	-aralo	-enla	-ose	

47 **terminations** = 4324 ciphers (Spanish) **Table 28.**

000	500	0000	5000	00000	50000	
←	←	←	←	←	←	
-abais	-ais	-arelo	-ela	-ennos	-ome	dollars
-abalo	-ales	-arete	-elos	-ese	-omelo	£ stg
-aban	-amele	-aria	-en	-eselo	-onos	frs
-abase	-anla	—	—	—	—	florins
-ada	-anlos	-arle	-enlas	-olas	-osele	marks
-adale	-anse	-arnos	—	—	—	barrels, casks or kegs *barrils, fûts etc.*
-adlas	-ar	-aron	—	—	—	bushels
-adlo	-arala	—	—	—	—	gallons
-adme	-arame	—	—	—	—	Kilos
-adola	-aras	-asen	-enlo	—	—	yards
-ados	—	—	—	—	—	tons of 20 cwt.
-adote	-areis	-aste	-enme	-olo	-otela	

Examples:
abrevales = 13 500 £ stg — argilome = 8 150 000 dollars — alistadote = 63 000

Table 29.

50 roots ×

General Sentences *Phrases Générales*

→		
egu–	it seems **advisable** to	*il semble à* ***propos*** *d(e)*
elij–	it does not seem advisable to	*il ne semble pas à propos d(e)*
enej–	we **are** (I am) about to	*nous sommes (je suis) en* ***train*** *d(e)*
enri–	we (I) **can**	*nous* ***pouvons*** *(je peux)*
entr–	„ „ „ not	*„ ne pouvons (je ne puis)*
erm–	„ „ „ probably	*„ pourrons (je pourrai) probablement*
evit–	„ „ „ „ not	*„ ne pourrons (je ne pourrai) probablement pas*
expi–	provided you can	*pourvu que vous puissiez*
fabe–	unless you can	*à moins que vous ne puissiez*
faj–	**endeavour** to	***tâchez*** *(de)*
fart–	it is absolutely **necessary** to	*il est absolument* ***nécessaire*** *d(e)*
fem–	we are (I am) **ready** to	*nous sommes (je suis)* ***prêt(s)*** *à*
ferr–	provided you are ready to	*pourvu que vous soyez prêt(s) à*
fili–	we (I) **recommend** you to	*nous (je) vous* ***recommand/ons(e)*** *d(e)*
firm–	„ „ „ „ not to	*„ „ „ „ de ne pas*
flet–	„ „ **request** you to	*„ vous* ***prions*** *(je v. prie) d(e)*
fogn–	„ „ „ „ not to	*„ „ „ „ „ „ de ne pas*
forj–	„ „ „ „ at once to	*„ „ „ „ „ „ d'immédiatement*
fos–	„ „ **shall**	*nous* ***allons*** *(je vais)*
fris–	„ „ „ not	*„ n'allons (je ne vais) pas*
frut–	„ „ „ probably	*„ allons (je vais) probablement*
garb–	„ „ „ „ not	*„ n'allons (je ne vais) probablement pas*
gay–	unless we (I) hear from you to the contrary by immed. telegr. reply we (I) shall	*sauf v. réponse contraire par télégramme immédiat nous allons (je vais)*
gib–	we (I) **wish** to	*nous* ***désirons*** *(je désire)*
gorm–	„ „ do not wish to	*nous ne désirons (je ne désire) pas*

wire ***télégraphiez***						
when *quand*	whether *si*	as soon as *aussitôt que*	to what extent *à quelle étendue*	on what terms *à quelles conditions*		
→	←	→	→	→		
gran–	had–	inco–	jet–	letr–	you can	*vous pouvez*
gru–	herb–	inst–	junt–	leud–	„ recommend to	*„ recommandez d(e)*
guip–	hurt–	jach–	—	—	„ intend to	*„ avez l'intention d(e)*
gust–	husm–	jadi–	labr–	licu–	„ are ready to	*„ êtes prêt/s à*
habl–	iman–	jam–	lact–	lill–	we are (I am) to	*nous devons (je dois)*
—	—	—	laqu–	—	and on what terms you can	*et à quelles conditions vous pouvez*
—	—	—	len–	—	and on what terms you recommend to	*et à quelles conditions vous recommandez d(e)*

65 **terminations** = 3250 ciphers (Spanish) **Table 29.**

←

-abais	accept	*accepter*	-arelas	limit	*limiter*
-abalas	„ draft	*„ traite*	-arelo	obtain	*obtenir*
-abalo	„ offer	*„ offre*	-arete	offer	*offrir*
-aban	act as proposed	*agir comme pro-*	-aria	pay	*payer*
-abanla	arrange	*arranger [posé*	-ariais	„ cheque	*„ chèque*
-abanme	buy	*acheter*	-arialo	„ draft	*„ traite*
-abanos	„ further	*„ davantage*	-ariase	prolong	*prolonger*
-abase	cancel	*annuler*	-arle	protest	*protester*
-ada	„ order	*„ ordre*	-arnos	quote	*coter*
-adale	cash	*encaisser*	-aron	reduce price	*réduire le prix*
-adlas	charter	*affrêter*	-aronla	„ prices	*„ les prix*
-adlo	conclude	*conclure [ché*	-aronse	remit	*remettre*
-adme	„ cheaper	*„ meilleur mar-*	-asemos	„ cheque	*„ chèque*
-adola	„ dearer	*„ plus cher*	-asen	„ 3 mo. s.	*„ 3 mois vue*
-adoles	cover	*couvrir*	-aste	renew contract	*renouveler contrat*
-adonos	deliver	*délivrer*	-asteis	repeat transact-	*répéter transac-*
-ados	„ within the time stipulated	*„ dans le délai stipulé*	-astela	replace [ion	*remplacer [tion*
			-asteme	sell	*vendre*
-adote	deposit	*déposer*	-ela	„ further	*„ davantage*
-ais	discount	*escompter*	-elos	„ in anticipa-	*„ à découvert*
-ales	do so	*faire celà*	-emonos	settle [tion	*régler*
-andolo	draw	*tirer*	-en	ship (send)	*expédier*
-andome	„ at sight	*„ à vue*	-enlas	„ „ further	*„ davantage*
-anla	„ 3 mo. s.	*„ „ 3 m. vue*	-enlo	„ „ in consignment	*„ en consignation*
-anlos	engage	*engager*			
-anse	„ freight	*„ nolis*	-enme	substitute	*substituer*
-ar	execute order	*exécuter ordre*	-ennos	supply	*fournir (= livrer)*
-arala	grant	*accorder*	-ese	take	*prendre*
-arales	„ credit	*„ crédit*	-olas	„ delivery of the goods	*„ livraison de la marchandise*
-arame	guarantee	*garantir*			
-aramos	„ quality	*„ qualité*	-olo	tender	*soumissionner*
-aranlo	increase price	*augmenter le prix*	-ome	wire information	*télégraphier renseignements*
-arante	„ prices	*„ les prix*			
-aras	insure	*assurer*	-onos	~~~~~~	~~~~~~
-areis	intervene	*intervenir*			

Examples:

evitaban = we (I) can probably not act as proposed. — *nous ne pourrons (je ne pourrai) probablement pas agir comme proposé.*

entradoles = we (I) cannot cover. — *nous ne pouvons (je ne puis) couvrir.*

Table 30.

18 roots ×

General Sentences *Phrases Générales*

lim-	~~~~~~	~~~~~~
lisi-	we (I) **have**	*nous* ***avons*** *(j'ai)*
logr-	„ „ „ as proposed	*„ „ „ comme proposé*
luch-	„ „ „ „ desired	*„ „ „ „ désiré*
ludi-	„ „ „ not	*nous n'avons (je n'ai) pas*
mago-	wire whether you have	*télégraphiez si vous avez*
malv-	we (I) understand that you have . . . ., if so, need not reply	*nous entendons (j'entends) que vous avez . . . . ., si oui, pas nécessaire répondre*
maul-	unless we (I) hear from you to the contrary by immediate telegraphic reply shall understand that you will have	*sauf v. réponse contraire par telégramme immédiat nous entendrons (j'entendrai) que vous aurez*
mazn-	when have you?	*quand avez-vous?*
med-	as soon as	*aussitôt*
mell-	wire as soon as	*télégraphiez aussitôt*
moj-	shall wire as soon as	*télégraphier/ons(ai) aussitôt*
mont-	wire to what extent you have	*télégraphiez à quelle étendue vous avez*
niel-	„ on what terms you have	*„ „ quelles conditions vous avez*
obvi-	if	*si*
ocup-	„ not	*„ pas*
odi-	not	*pas*
ojal-	„ yet	*„ encore*

Examples:

malvabalo = we (I) underst. that you have accepted offer, if so, need not reply.
nous (j') entend/ons(s) que vous avez accepté offre, si oui, pas néc. rép.

maznasemos = when have you remitted cheque?
quand avez-vous remis chèque?

65 **terminations** = 1170 ciphers (Spanish) **Table 30.**

-abais	accepted	*accepté*
-abalas	„ draft	„ *traite*
-abalo	„ offer	„ *offre*
-aban	acted as proposed	*agi comme proposé*
-abanla	arranged	*arrangé*
-abanme	bought	*acheté*
-abanos	cancelled	*annulé*
-abase	„ order	„ *ordre*
-ada	cashed	*encaissé*
-adale	chartered	*affrêté*
-adlas	concluded	*conclu* [*ché*
-adlo	„ cheaper	„ *meilleur mar-*
-adme	„ dearer	„ *plus cher*
-adola	covered	*couvert*
-adoles	delivered	*délivré*
-adonos	deposited	*déposé*
-ados	discounted	*escompté*
-adote	done so	*fait celà*
-ais	drawn	*tiré*
-ales	„ at sight	„ *à vue*
-andolo	„ 30 days's.	„ *30 jours vue*
-andome	„ 3 mo. s.	„ *3 mois vue*
-anla	engaged	*engagé*
-anlos	„ freight	„ *nolis*
-anse	executed order	*exécuté ordre*
-ar	„ all orders	„ *tous les ordres*
-arala	granted	*accordé*
-arales	„ credit	„ *crédit*
-arame	guaranteed	*garanti*
-aramos	increased price	*augmenté le prix*
-aranlo	„ prices	„ *les prix*
-arante	insured	*assuré*
-aras	limited	*limité*
-areis	obtained	*obtenu*
-arelas	offered	*offert*

-arelo	ordered	*commandé*
-arete	paid	*payé*
-aria	„ cheque	„ *chèque*
-ariais	„ draft	„ *traite*
-arialo	prolonged	*prolongé*
-ariase	protested	*protesté*
-arle	quoted	*coté*
-arnos	received	*reçu*
-aron	reduced price	*réduit le prix*
-aronla	„ prices	„ *les* „
-aronse	remitted	*remis*
-asemos	„ cheque	„ *chèque*
-asen	„ 30 days's.	„ *30 jours vue*
-aste	„ 3 mo. s.	„ *3 mois* „
-asteis	{ „ amount of Invoice	{ „ *montant de facture*
-astela	renewed contract	*renouvelé contrat*
-asteme	{ repeated transaction	*répété transaction*
-ela	replaced	*remplacé*
-elos	sold	*vendu*
-emonos	„ in anticipa-	„ *à découvert*
-en	settled [tion	*réglé*
-enlas	shipped (sent)	*expédié*
-enlo	{ „ „ in consignment	{ „ *en consignation*
-enme	stipulated	*stipulé*
-ennos	substituted	*substitué*
-ese	taken	*pris*
-olas	{ „ delivery of the goods	{ „ *livraison de la marchandise*
-olo	tendered	*soumissionné*
-ome	wired information	{ *télégraphié renseignements*
-onos	~~~~~~	~~~~~~

Table 31.

24 roots ×

Shipments *Expéditions.*

oliv–	ship (send)	*expédiez*
oper–	„ „ only	*„ seulement*
opil–	continue to ship (send)	*continuez à expédier*
orin–	do not ship (send)	*n'expédiez pas*
oxid–	„ „ continue to ship (send)	*ne continuez pas à expédier*
pali–	it seems advisable to ship (send)	*il semble à propos d'expédier*
pasm–	„ „ „ not to ship (send)	*„ „ „ „ de ne pas expédier*
paut–	provided you ship (send)	*pourvu que vous expédiez*
pect–	unless „ „ „	*à moins que vous n'expédiez*
pein–	we (I) shall ship (send)	*nous expédierons (j'expédierai)*
peor–	„ „ can „ „	*nous pouvons (je peux) expédier*
pist–	„ „ „ continue to ship (send)	*„ „ „ „ continuer à expédier*
plom–	it is impossible to ship (send)	*il est impossible d'expédier*
puj–	„ „ „ to continue shipping (sending)	*„ „ „ de continuer à expédier*
quem–	we (I) can probably ship (send)	*nous pourrons (je pourrai) probablement expédier*
rasp–	„ „ „ „ not „ „	*nous ne pourrons (je ne pourrai) probablement expédier*
rear–	wire whether you can ship (send)	*télégraphiez si vous pouvez expédier*
regl–	provided you can ship (send)	*pourvu que vous puissiez expédier*
rend–	unless „ „ „ „	*à moins que vous ne puissiez expédier*
rest–	wire whether we are (I am) to ship (send)	*télégraphiez si nous devons (je dois) expédier*
ripi–	wire whether you recommend shipping (sending)	*télégraphiez si vous recommandez d'expédier*
roci–	shipment (dispatch)	*expédition*
roll–	instead of shipping (sending)	*au lieu d'expédier*
ruin–	shipped (sent)	*expédié*

Examples:

reglandome = provided you can ship (send) by steamer without transshipment / *pourvu que v. puissiez exp. p. vap. sans transbordement.*

o inanla = do not ship by another steamer / *n'expédiez pas par un autre vapeur*

←

-abais	～～～～	～～～～
-abalas	against contract	*à valoir sur contrat*
-abalo	in consignment	*en consignation*
-aban	on trial	*à titre d'essai*
-abanla	total **quantity**	***quantité** totale*
-abanos	„ „ available	*„ „ disponible*
-abase	the quantity mentioned	*la quantité mentionnée*
-ada	half quantity	*moitié de la quantité*
-adale	double „	*double quantité*
-adlas	large „	*grande „*
-adlo	small „	*petite „*
-adme	larger „	*quantité plus grande*
-adola	smaller „	*„ „ petite*
-adoles	further	*davantage*
-adonos	the remainder	*le reste*
-ados	part only	*partie seulement*
-adote	goods ordered	*marchandise(s) commandée(s)*
-ais	in the **manner** mentioned	*de la **manière** mentionnée*
-ales	by steamer	*par vapeur*
-andolo	„ „ transshipment at	*„ „ transbordement à*
-andome	„ „ without transshipment	*„ „ sans transbordement*
-anla	„ another steamer	*„ un autre vapeur*
-anlos	„ direct „	*„ vapeur direct*
-anse	„ first steamer leaving	*„ premier vapeur en partance*
-ar	„ sailer	*„ voilier*
-arala	„ cheapest route	*„ voie la moins chère*
-arales	„ the usual line	*„ la ligne habituelle*
-arame	usual **quality**	***qualité** habituelle*
-aramos	superior „	*„ supérieure*
-aranlo	current „	*„ courante*
-arante	cheaper „	*„ meilleur marché*
-aras	dearer „	*„ plus chère*
-areis	quality = to sample	*„ = à échantillon*
-arelas	„ = to your sample	*„ = à votre échantillon*
-arelo	„ about = to sample	*„ à peu près = à échantillon*
-arete	same quality as last time	*même qualité que la fois dernière*
-aria	{ „ „ and on same terms as last time	*{ „ „ et aux mêmes conditions que la fois dernière*
-ariais	the best quality procurable	*la meilleure qualité procurable*
-arialo	quality guaranteed	*qualité garantie*
-ariase	„ „ = to	*„ „ = à*
-arle	the quality mentioned	*la qualité mentionnée*
-arnos	on the **terms** mentioned	*aux **conditions** mentionnées*
-aron	„ „ usual terms	*„ „ habituelles*
-aronla	at lower freight	*à frêt plus bas*
-aronse	„ higher „	*„ „ „ élevé*
-asemos	„ chance „	*„ „ d'occasion*
-asen	„ through rate	*„ taux à forfait*
-aste	within the **time** mentioned	*dans le **délai** mentionné*
-asteis	immediately	*immédiatement*
-astela	within 8 days	*dans les 8 jours*
-asteme	this week	*cette semaine*
-ela	next „	*semaine prochaine*
-elos	end of this week	*fin de cette semaine*
-emonos	„ „ next „	*„ „ semaine prochaine*
-en	beginning of next week	*commencement de semaine prochaine*
-enlas	this month	*mois courant*
-enlo	next „	*„ prochain*
-enme	this and next month	*„ courant et prochain*
-ennos	next 2 months	*les 2 mois prochains*
-ese	2nd half of current month	*2e moitié mois courant*
-olas	1st „ „ next „	*1ère „ „ prochain*
-olo	2nd „ „ „ „	*2e „ „ „*
-ome	weekly	*par semaine*
-onos	monthly	*mensuellement*

ship&po.
31

Table 32. Ports &c. 15 roots ×

➛

sain–	fob	*franco bord*
silb–	c. & f.	*c. & f.*
sis–	cif	*c. f. & a.*
siti–	steamer to	*vapeur à destination d/e*
soas–	sailer „	*voilier „ „ „*
sob–	freight by steamer to	*fret par vapeur à destination d/e*
sond–	from	*d/e*
sopl–	option	*option*
sufl–	and/or	*et/ou*
tap–	via	*via*
tarj–	to be shipped to	*à expédier à*
tech–	ship to	*expédiez à*
ters–	we (I) have shipped to	*nous avons (j'ai) expédié à*
tir–	shipment to	*expédition à (à destination d/e)*
tizn–	~~~~	~~~~

Examples:

sainabais = fob Aberdeen — *franco bord Aberdeen*

soplarle = option Dunkirk — *option Dunkerque*

Table 33. Ports &c. 15 roots ×

➛

told–	fob	*franco bord*
tom–	c. & f.	*c. & f.*
torn–	cif	*c. f. & a.*
trab–	steamer to	*vapeur à destination d/e*
trep–	sailer „	*voilier „ „ „*
truf–	freight by steamer to	*fret par vapeur à destination d/e*
tumb–	from	*d/e*
uvi–	option	*option*
vaci–	and/or	*et/ou*
valu–	via	*via*
vigi–	to be shipped to	*à expédier à*
vot–	ship to	*expédiez à*
zanj–	we (I) have shipped to	*nous avons (j'ai) expédié à*
zarp–	shipment to	*expédition à (à destination d/e)*
zurr–	~~~~	~~~~

Examples:

toldabais = fob Liverpool — *franco bord Liverpool*

65 **terminations** = 975 ciphers (Spanish). **Table 32.**

←

-abais	Aberdeen
-abalas	Adelaide
-abalo	Aden
-aban	Alexandria
-abanla	Amsterdam
-abanme	Antwerp / *Anvers*
-abanos	Auckland
-abase	Bahia
-ada	Baltimore
-adale	Barcelona
-adlas	Batoum
-adlo	Batavia
-adme	Belfast
-adola	Black See / *Mer Noire*
-adoles	Bluff Harbour
-adonos	Bombay
-ados	Bordeaux
-adote	Boston
-ais	Bremerhaven
-ales	Brindisi
-andolo	Brisbane
-andome	Bristol
-anla	Buenos Aires
-anlos	Calcutta
-anse	Canal
-ar	„ for ordres / „ *à ordres*
-arala	Cape
-arales	„ Good Hope
-arame	Capetown
-aramos	Cardiff
-aranlo	Chittagong
-arante	Christiania
-aras	Civita Vecchia
-areis	Colombo (Ceylon)
-arelas	Constantinople
-arelo	Continent
-arete	„ for orders / „ *à ordres*
-aria	Copenhagen
-ariais	Delagoa Bay
-arialo	Dublin
-ariase	Dundee
-arle	Dunkirk *(Dunker-[que)*
-arnos	Fiume
-aron	Genoa / *Gênes*
-aronla	Gibraltar
-aronse	Glasgow
-asemos	Grimsby
-asen	Haiphong
-aste	Hamburg
-asteis	Hartlepool (West)
-astela	Havana
-asteme	Hâvre
-ela	Hiogo
-elos	Hongkong
-emonos	Hull
-en	Ibraïl
-enlas	Iquique
-enlo	Ivory Coast / *Côte d'Ivoire*
-enme	Kurrachee
-ennos	La Plata
-ese	Las Palmas
-olas	Leeds
-olo	Leghorn / *Livourne*
-ome	Levant ports / *ports de Levante*
-onos	Lisbon / *Lisbonne*

---

65 **terminations** = 975 ciphers (Spanish) **Table 33.**

←

-abais	Liverpool
-abalas	London
-abalo	Los Angeles
-aban	Madras
-abanla	Manila
-abanme	Marseille/s
-abanos	Mediterranean / *Méditerranée*
-abase	Melbourne
-ada	Montevideo
-adale	Nagasaki
-adlas	Naples
-adlo	New Brunswick
-adme	„ Orleans
-adola	„ York
-adoles	Newcastle
-adonos	Odessa
-ados	Panama
-adote	Penang
-ais	Philadelphia
-ales	Port Bou
-andolo	„ Elizabeth
-andome	„ Natal
-anla	„ Philip
-anlos	Port Pirée
-anse	Port Said
-ar	Porto
-arala	Porto Alegre
-arales	Puerto Principe
-arame	Rangoon
-aramos	Red Sea
-aranlo	Riga
-arante	Rio Janeiro
-aras	„ Grande do Sul
-areis	Rosario
-arelas	Rotterdam
-arelo	Saigon
-arete	San Francisco
-aria	St. Johns N. F.
-ariais	„ Louis
-arialo	„ Petersburg
-ariase	„ Vincent C. V.
-arle	Santa Rosalia
-arnos	Santos
-aron	Scarborough
-aronla	Shanghai
-aronse	Singapore
-asemos	Southampton
-asen	Straits Settl.
-aste	Stettin
-asteis	Sulina
-astela	Sydney
-asteme	Table Bay
-ela	Tientsin
-elos	Trieste
-emonos	U. K. / *Roy. Uni*
-en	U. K. or Continent / *Roy. Uni ou Cont.*
-enlas	U. K. or Cont. for orders / *R. U. ou Cont. à ordres*
-enlo	Valparaiso
-enme	Vancouver
-ennos	Vladivostok
-ese	Yokohama
-olas	Zansibar
-olo	your port *(votre port)*
-ome	our „ *(notre port)*
-onos	direct port *(port direct)*

Table 34 Terms *Conditions* 50 roots ×

→		
abol–	~~~~~~	~~~~~~
acut–	net	*net*
alid–	„ per bag	*„ par sac*
annu–	„ „ bale	*„ „ balle*
aper–	„ „ barrel (cask, keg)	*„ „ baril (fût etc.)*
brun–	„ „ bushel	*„ „ bushel*
carp–	„ „ case	*„ „ caisse*
cest–	„ „ „ of 1 doz	*„ „ „ de 1 douzaine*
civ–	„ „ „ „ 25 small bottles	*„ „ „ „ 25 petites bouteilles*
comp–	„ „ „ „ 50 „ „	*„ „ „ „ 50 „ „*
cond–	„ „ „ „ 25 large „	*„ „ „ „ 25 grandes „*
covr–	„ „ „ „ 50 „ „	*„ „ „ „ 50 „ „*
cusc–	„ „ cwt	*„ „ cwt*
dolc–	„ „ „ gross weight	*„ „ „ poids brut*
esib–	„ „ „ net weight	*„ „ „ „ net*
fed–	„ „ dozen	*„ „ douzaine*
fin–	„ „ „ pints	*„ „ „ pints*
fior–	„ „ „ quarts	*„ „ „ quarts*
fogl–	„ „ gallon	*„ „ gallon*
forb–	„ „ foot (engl.)	*„ „ pied (angl.)*
fugg–	„ „ square foot (engl.)	*„ „ „ carré (angl.)*
fulc–	„ „ cubic foot (engl.)	*„ „ „ cube (angl.)*
graz–	„ „ gross	*„ „ grosse*
grem–	„ „ kilo	*„ „ Kilo*
guar–	„ „ 100 kilo	*„ „ 100 Kilos*
inib–	„ „ 1000 kilo	*„ „ 1000 Kilos*
inve–	„ „ 100 liter	*„ „ 100 litres*
lald–	„ „ meter	*„ „ mètre*
larg–	„ „ „ square	*„ „* m^2
malt–	„ „ cubicmeter	*„ „* m^3
minu–	„ „ oz (engl.)	*„ „ once (angl.)*
mun–	„ „ piece	*„ „ pièce*
nodr–	„ „ 100 pieces	*„ „ 100 pièces*
nutr–	„ „ 1000 „	*„ „ 1000 „*
ont–	„ „ lb (engl.)	*„ „ lb (angl.)*
paur–	„ „ „ „ net weight	*„ „ „ „ poids net*
pol–	„ „ „ „ gross weight	*„ „ „ „ „ brut*
relu–	„ „ quarter (28 lbs)	*„ „ quarter (28 lbs)*
riun–	„ „ ton of 20 cwt	*„ „ tonne de 20 cwt*
sanc–	„ „ „ „ „ „ net weight	*„ „ „ „ „ „ poids net*
sbad–	„ „ „ „ „ „ gross weight	*„ „ „ „ „ „ „ brut*
scur–	„ „ „ „ 2000 lbs	*„ „ „ „ 2000 lbs*
sent–	„ „ yard	*„ „ yard*
serr–	„ „ square yard	*„ „ „ carré*
sfug–	„ „ cubic yard	*„ „ „ cube*
sped–	from this moment, unless otherwise expressed, all prices **for this firm** will be understood	*dès ce moment, à moins d'indications contraires, tous les prix **pour cette maison** s'entendront*
stru–	from this moment, unless otherwise expressed, all prices **of this article** will be understood	*dès ce moment, à moins d'indications contraires, tous les prix **de cet article** s'entendront*
tort–	from this moment, unless otherwise expressed, all prices will be understood	*dès ce moment, à moins d'indications contraires, tous les prix s'entendront*
trad–		
vom–		

fob		c. & f.		cif *(c. f. & a.)*		
our (this) port *notre (ce) port*	**your** port *votre port*	**our** (this) port *notre (ce) port*	**your** port *votre port*	**our** (this) port *notre (ce) port*	**your** port *votre port*	
←	←	←	←	←	←	
-endo	-iamovi	-irai	-iresti	-irollo	-issero	~~~~~~
-endolo	-icci	-iranle	-irete	-irommi	-istelo	usual payment terms *conditions habituelles de payement*
-endone	-illo	-iranno	-irla	-ironla	-istene	net cash against documents *net comptant contre documents*
-endovi	-immo	-iransi	-irmelo	-irono	-isti	net 3 months against Letter of Credit *net 3 mois contre lettre de crédit*
-ente	-immolo	-ircela	-irmene	-irossi	-istici	net 3 months against documentary credit *net 3 mois contre crédit documentaire*
-entesi	-immone	-irebbe	-irne	-irotti	-istila	net 3 months or 2 % 30 days *net 3 mois ou 2 % 30 jours*
-iamolo	-immoti	-iremo	-iro	-isse	-itala	net 3 months or $2\frac{1}{2}$ % 30 days *net 3 mois ou $2\frac{1}{2}$ % 30 jours*

←		
-itami	free	*franco*
-ite	„ on railway cars	„ *sur wagon*
-iteci	„ „ „ „ this place	„ „ „ *cette place*
-itele	„ „ „ „ your „	„ „ „ *votre* „
-itivi	„ of railway carriage and import duty	„ *de port (par chem. de fer) et droits d'entrée*
-itolo	„ „ „ „ „ „ „ this place	*franco de port (par chem. de fer) et droits d'entrée cette place*
-itone	free of railway carriage and import duty your place	*franco de port (par chem. de fer) et droits d'entrée votre place*
-itosi	free house (works) this place	*franco domicile (usine/s) cette place*
-iva	„ „ „ your „	„ „ „ *votre* „
-ivaci	payment against delivery	*payement contre livraison*
-ivale	„ „ „ free on railway cars this place	„ „ „ *franco wagon cette place*
-ivanla	payment against delivery free on railway cars your place	*payement contre livraison franco wagon votre place*
-ivano	free quay	*franco quai*
-ivanti	„ „ this place	„ „ *cette place*
-ivilo	„ „ your „	„ „ *votre* „
-ivimi	inclusive of packing	*emballage compris*
-ivine	packing extra	„ *non compris*
-ivola	~~~~~~	~~~~~~
-ivoti		
-ivvi		

Examples:

grazistelo = net p. gross cif your port usual payment terms
net p. grosse c. f. & a. v. port, cond. hab. de payement

riunivanti = net p. ton of 20 cwt free quay this place
net p. tonne de 20 cwt franco quai cette place

Table 35 Private *Privée.* 41 roots ×

→

abdic–
ableg–
abloc–
abnod–
accur–
adaer–
adapt–
adjug–
agger–
alien–
allig–
ambul–
amput–
anhel–
annot–
appar–
ariet–
arrog–
aspir–
aucup–
augur–
aurig–
autum–
brevi–
calce–
calig–
castr–
celer–
clipe–
coept–
color–
comit–
comme–
confl–
coron–
cribr–
cruci–
crust–
dealb–
deaur–
delig–

37 terminations = 1517 ciphers (Latin) **Table 35.**

←

-abam
-abant
-abare
-abis
-abo
-amini
-amur
-andae
-andi
-andos
-andum
-ans
-antem
-antis
-anto
-antur
-arent
-arere
-ares
-ari
-arier
-at
-atae
-atior
-atius
-atos
-atote
-atum
-aturi
-avero
-avit
-emus
-ent
-es
-etis
-etur
-or

Priv.

Table 36 Private *Privée.* 172 roots ×

	→	→	→	→
	avisp– azolv– azorr– azufr– baraj–	desun– devan– deved– devis– dibuj–	escot– escud– esfri– esfum– esmer–	irrit– istri– jabel– jambr– jarop–
	barlo– batoj– betun– blasm– bolet–	difam– disen– disip– domen– drans–	esmol– espad– espej– espit– estib–	jorob– junel– lamin– lembr– limpi–
	boved– brind– buril– calon– cambi–	embaf– embij– embud– empaj– empil–	estof– exalt– execr– exent– exhum–	madur– manch– manej– maran– marid–
	camel– camod– carem– cavil– cecin–	empop– empun– enagu– enalb– enart–	exorn– fabul– fatur– filtr– flech–	mayor– menor– mingu– model– moned–
	chant– charp– chipr– chorr– churn–	enasp– encaj– encep– encim– endon–	forci– franj– freil– fusil– galib–	motej– murci– musit– mutil– nervi–
	cobij– cogit– cohob– comis– conlo–	enfil– engaf– engib– engom– enhad–	gerib– gradu– guard– gueld– guind–	novel– objet– obtur– ocult– ofens–
	copel– copin– corch– crism– cuart–	enhen– enhot– enjeb– enjut– enlam–	hacin– hered– honor– horad– imput–	ofici– olvid– ondul– opres– opugn–
	cuest– dadiv– dehes– delax– demor–	enlos– enrun– ensan– ensop– entes–	incub– inhal– inici– inmol– inmut–	orden– orill– pavon– pinch– pinor–
	demud– depos– desoj–	envin– equip– escap–	innov– inund– invej–	preci– prens– puntu–

47 **terminations** = 8084 ciphers (Spanish) **Table 36.**

←

-abais
-abalo
-aban
-abase
-ada
-adale
-adlas
-adlo
-adme
-adola
-ados
-adote
-ais
-ales
-amele
-anla
-anlos
-anse
-ar
-arala
-arame
-aras
-areis
-arelo
-arete
-aria
-arle
-arnos
-aron
-asen
-aste
-ela
-elos
-en
-enlas
-enlo
-enme
-ennos
-ese
-eselo
-olas
-olo
-ome
-omelo
-onos
-osele
-otela

Index of Text

The figures not preceded by a "p" are the table numbers.

a

about to 29
accept 1 & 2, 17/9 (quant.), 29 & 30, p. 1, p. 17 (prop.)
acceptance p. 1
accident p. 1
account p. 1
act 1, 29 & 30, p. 1
address p. 1
advice p. 1
advisable 29, p. 1
afloat 14
agree 1, p. 1
allowance p. 1, p. 18 (qual.), p. 20 (ship.), p. 22 (weight)
aloof p. 9 (hold)
amount/s 12, 24/8, p. 1
&/or 32/3 (ports)
appeal p. 1
appear p. 1
approval p. 1, p. 4 (consent)
arbitration p. 1
area p. 5 (crop)
arrange 29 & 30, p. 1
arrival (date) 4 & p. 5, 7 (hours), 20, p. 1
arrivals 14
as soon as 29/30, p. 20
assets p. 1
auction 20, p. 1
available 31, p. 1, p. 22 (unav.)
average 16, p. 1

b

bags 18, 26/7, 34(terms), p. 1
balance 11 (finance), p. 1
bales 18, 26/7, 34 (per)
ballast 4 (dates), p. 1
barrels 18, 26/8, 34(per), p. 1
barsilver 13
be p. 1
before 4(dates), 5(Epo)
believe p. 2
best 1 (act), p. 2
bill p. 2, p. 4 (coll.), p. 6 (disc.), p. 9 (health), p. 18 (protest)
B/L p. 2, p. 5 (dated)
boxes p. 2
brokerage p. 2
bushel/s 26, 27/8
buy 17/9 (quant.), 20, 21/3, 29, 30, p. 2, p. 6 (against)
buyers 20/2, 23 (±), p. 2, p. 12 (market)

c

can 29, p. 3
cancel 4 (dates), 29, 30, p. 3, p. 12 (limit), p. 13 (offer), p. 14 (order)
capital 9, 10, p. 3, p. 14 (overdr.)
care p. 3, p. 1 (addr.)
cargo p. 3, p. 5 (damaged), p. 6 (deck), p. 11 (jettisoned)
carriage 34(terms), p. 3
cases 18, 26/7, 34 (per)
cash 29, 30, 34 (terms), p. 3, p. 15 (payment)
cask/s 18, 26/8, 34
centimes 26
centimetres 26
cents 24 & 26
chance 31 (freight), p. 3, p. 2 (buying), p. 9(frt), p. 13(offer), p. 16 (price), p. 17 (prospects), p. 19 (sale)
charter 29, 30, p. 3
cheaper 23
cheque 29/30 (pay & remit), p. 3
chests p. 3
class p. 3, p. 8 (fare), p. 15 (passage)
clear p. 3
clock 7, 8
code p. 3 & 4
collection p. 2 (bill), p. 4, p. 7 (draft)
communicate p. 4
compete p. 4
competence 9 & 10
competition p. 4
competitors 20, 23(±)
conclude 17/9, 23 (±), 29, 30
condition p. 4, p. 9 (health)
confirmation 1, p. 4
conflicting p. 19 (reports)
consent p. 4, p. 1 (approval)
consign 31, p. 4
consignment 29/31 (ship), p. 4
Consul p. 4
consumption 14, p. 4
contract 29 & 30 (renew), 31, p. 4 & 5
copy p. 5
correct 1 & 2, p. 5
correspondence 1, p. 5
c & f, c. i. f. 20, 32/4
cost p. 5, p. 6 (dism.)
Court p. 5, p. 11 (judgm.)
cover 20/1, 29, 30
credit 9 & 10 (inf.), 11, 29 & 30 (grant), 34, p. 5
creditors p. 5, p. 13 (meet.)
crew p. 5
crop 15/6, 20/2, p. 5
custom p. 5, p. 7 (duty)
cwt 19, 26, 34
cyclone 15 (crop), p. 5

d

d. 23 (±), 24, 26
damage 15 & 21 (crop), p. 5, p. 10 (injured)
date 3, 4, p. 5, p. 3 (cancel), p. 7 (due)
day 7, 26, p. 5, p. 11 (laydays)
deadweight p. 5, p. 9 (frt.)
dearer 23 (±), 29, 30
death p. 5, p. 6 (died)
debt p. 9 (floating)
declared p. 6, p. 22 (value)
declining 13, 22
decreasing 14, 22
delay 4 (dates), p. 6
delivery 4 (date), 5 & 6 (Epo), 29/30 (del. & take), p. 6
demand 14, 21/2, p. 6
departure 4, p. 5 (date), p. 6, p. 11 (leave)
deposit 29 & 30, p. 6
destroyed 15 (crop), p. 6
details p. 6, p. 14 (part.)
died p. 6, p. 5 (death)
difficulties 9 (inf.) p. 6
discount 13, 29, 30, 34, p. 6
dispatch (ship, send) 1 (order), 3 & 4 (date), 5 & 6 (Epo), 17/9 (quant.), 31
disposal p. 6
dividend p. 6
do 29, 30, p. 6
document/s 34 (terms), p. 6
dollar/s 12, 25/8
double quant. 20 & 31, p. 18
doz. 26, 34 (per)
drafts 11, 12, 29 & 30, (accept & pay), p. 7
drought 15 (crop)
due 3 (date), p. 7
during 5 (Epo)
duty 34, p. 7

e

endeavour 29, p. 7
engage 29, 30
engagements p. 7
error 1 & 2 (mistake), p. 7, p. 13 (mistake)
established 9 & 10, p. 7
exchange 11, 13, p. 7/8
execute 29 & 30, p. 14 (order)
exempt p. 8, p. 9 (free)
expect p. 8, p. 4 (conf.)
exporters 9 & 10
extend offer 4
extension 4 (date), p 8
extent 29, 30, p. 8

f

failure 9, p. 8, p. 14 (orders)
falling 13, 22
feet 26
finance 9/12, p. 8
find p. 9
finish p. 8
fire p. 8
firm p. 8
florin 26/8
fluctuations 13
fob 20, 32 3, 34, p. 9
foot 34 (per), p. 9
force 4 (date), p. 9
fractions 4 (dates), 24/6
francs 12, 25, 27/8
free 34, p. 9
freight/s 13, 14 (tonn.), 29 & 30, (engage), 31 (ship.), 32/3 (ports), p. 9, p. 21 (through), p. 22 (tonn.)
frost 15 (crop)
funds 9 & 10, p. 9

g

gallon/s 26/8, 34 (per)
grant 29, 30
gross 26, 34
gross wt 34, p. 22/3 (wt)
guarantee 29, 30, p. 4 (comm.), p. 9

h

half 4 (dates), 20, 26, 31 (ship)
hand p. 9, p. 2 (buy)
health p. 9, p. 10 (illness)
heat 15 (crop)
high/er 13, 22/3
hours 7 & 8

i

if 30, p. 10, p. 18 (prov.)
illness p. 10, p. 9 (health)
immovable p. 10, p. 17 (prop.)
importers 9 & 10
increase 14, 21, 29 & 30 (prices), p. 10
information 9, 10, 29 & 30 (wire), p. 10, p. 13 (news), p. 19 (reports)
inquire 9, p. 10
insects 15 (crop)
insolvency p. 10, p. 14 (orders)
instalments p. 10, p. 15 (pay)
instructions p. 10
insure 29, 30, p. 10
intend 29, p. 10
interest p. 10, p. 15 (pay)
intervene 29
Invoice 3, 11, 30 (remit), p. 11, p. 5 (copy)
involved p. 11, p. 8 (failed)

k

kegs 18, 26/8
kilo 19, 26/8, 34 (per)
know p. 11, p. 8 (firm)

l

late p. 11, p. 6 (deliv.), p. 20 (shipm.)
later 4 (dates), 5 (Epo)
laydays p. 11
leave 4 (date), 7 (hours), p. 11
legal p. 11, p. 17 (proc.)
legalized p. 3 (certific.)
letter 1, 3, 11 (credit), p. 5 (cred.), p. 11
liabilities p. 11, p. 1 (assets), p. 8 (failed)
liable p. 11, p. 19 (respons.)
limit 20/3, 29, 30, p. 11/2
line 31
Lire it 27
liter 34 (per)
load 4 (dates), p. 12
locusts 15
loss p. 12, p. 22 (weight)
losses 9
lost p. 2 (board), p. 12
Lstg 12, 24, 26/8
low/er 13, 21, 23

m

make p. 12
manufacturer/s 9 & 10, p. 12 (makers)
margin 23, p. 12
mks 12, 25/8
market 13, 23 (±), p. 12/3, p. 7 (estim.)
maximum 17/9, p. 13, p. 18
meaning 2, p. 13
means 9 & 10, p. 13
measures 17, 26/8
meeting (credit.) p. 13
meters 17, 26, 34 (per)
m/m 26
milreis 27
mistake 1, 2, p. 13, p. 7 (error)
money 24/8, p. 13
months 5, 6, 20 (opera), 31 (ship)
monthly 31 (ship)
morality 10
movable p. 17 (property)
movements 13, 14

n

necessary 29, p. 13
net 34 (per), p. 22/3 (weight)
news p. 13, p. 10 (inform.), p. 19 (reports)
nominal 13, p. 13, p. 17 (prices), p. 18 (quotns)

o

obtain 29, 30
offer 1 (tel. & lett.), 2 (tel.), 4 (dates), 6 (Epo), 7 (hours), 17/9 (quant.) 20/3, 29/30, p. 13
offerings 14
on (Ldn. &c.) 11
opinion p. 13/4
option 23 (±), 32/3 (ports), p. 3 (to cancel), p. 14
order 1, 3, 29 & 30 (cancel & execute), 31 (ship), 32/3 (ports) p. 14, p. 6 (delivery)
ordering 30
oz 26, 34 (per)
overdrawn p. 14, p. 5 (cred.)
oversold p. 15 (pos.)

p

packing 34 (terms), p. 14
parcel p. 14, p. 12 (lot)
particulars p. 14, p. 6 (details)
passage p. 14/5
passenger p. 15, p. 5 (crew)
payable 3 (date), p. 15
payment 10 (inf.), 12, 29, 30, 34 (terms), p. 15, p. 16 (prepaym.)
pence 23 (±), 24, 26
per cent 23 (±), 26, p. 13 (more)
Pfg 26
piece 34 (per)
pints 34 (p. doz)
place p. 15
places 32 3
ports 32/3, p. 15
position 9, 10, p. 8 (financ.), p. 15, p. 14 (orders)
possible p. 15, p. 2 (better), p. 7 (early), p. 13 (much)
post p. 15, p. 14 (parcel)
£ stg 12, 24, 26/8
lb/s 19, 26, 34 (per)
Prices 13 (movem.), 17/9 (quant), 20, 21 & 22 (present & future), 23 (±), 24/6, 29/30 (increase & reduce), 34 (understood), p. 16/7, p. 20 (sample)
production 14 (mov.), 16 (stat.), 21/2 (opera) p. 17
prolong 29, 30
proposed, 29 & 30 (act), p. 17
prospects p. 17, p. 3 (chances)
protest 29, 30, p. 17/8

q

quality 13 & 14 (mov.), 16 (stat.), 20 (opera), 29 (guarantee), 31 (ship), p. 18, p. 16/7 (prices), p. 20 (sample)
quantity 20 (opera), 29 & 30 (extent), 31 (ship), p. 18
quantities 17/9, 24/8
quarter 4 (dates), 5 (Epo)
qr/s 26/7, 34 (per)
quarts 34 (per doz)
quay 34 (terms), p. 18
quotations 13, p. 18
quote 29 & 30

r

railway cars 34 (terms)
rain 15 (crop), p. 18
ready 29, p. 18

realization 22, p. 18, p. 19 (sales)
receipts 14 (mov.), 21/2 (opera), p. 18
receive 1 (tel. & lett.), 12 (fin.), 30
recommend 29
reduce 29 & 30 (prices), p. 19
refusal p. 19, p. 20 (sample)
remainder 4 (date), 20 (quant.), 31 (ship)
remittances 11/2, 29, 30, p. 3 (cashed), p. 19
renew 1, 7 & 23 (offer), 29, 30, p. 8 (exhausted), p. 19
repeat 2 (ciphers), 29 & 30 (transact.), p. 19
replace 29, 30, p. 8 (exhausted), p. 19
reply 1 (lett. & tel.), 2 (tel.), 7 (hours), p. 19, p. 21 (tel.)
reports p. 19, p. 10 (inform.), p. 13 (news)
request 29, p. 19
responsible p. 19, p. 11 (liable)
result p. 19
retail p. 19
retire p. 8 (firm)
risk p. 19, p. 8 (expense)
rising 13 (mov.), 21 (opera.)
route 31 (ship.), p. 19
rubles 27
rumours p. 19, p. 15 (polit.)
rupees 27

s

saved p. 2 (board)
sales 17/9 (quant.), 20 (opera), 29, 30, p. 1 (auction), p. 19/20, p. 11 (limit)
samples 16 (qual.), 20 & 31 (qual.), p. 20, p. 18 (qual.)
season 5 (Epo)
sellers 20/2 (opera), 23 (±), p. 2 (buyers), p. 12 (mkt), p. 22 (unavail.)
send 1 (order), 3 & 4 (dates), 17/9 (quant.), 29/31.
settle 29, 30
settlement p. 20, p. 16/7 (price)
shilling p. 26
shipment 1, 3 & 4 (dates), 5 & 6 (Epo), 14 (mov.), 16 (stat.), 17/9 (quant.), 20/2 (opera), 29/31, 32/3 (ports), p. 20, p. 4 (completed), p. 11 (late)
shipbroker/s 9, 10
sight 11, p. 20
silver 13
solicit 29 & 30
soon 29 & 30 (as soon as), p. 20
spare p. 20, p. 3 (cash)
specul^{n} 9, p. 20/1, p. 8 (failed)
spot 20 (opera), p. 21
spring 5
square 26, 34 (per)
standing 9, 10, p. 21, p. 15 (pos.)
state p. 21, p. 4 (condition), p. 9 (health)
statistics 15/6
steamer 31 (ship), 32/3 (ports), p. 21
stipulate 30, p. 21
stock/s 14 (mov.), 21/2 (opera), p. 21, p. 6 (demand), p. 17 (prices)
substitute 29, 30, p. 21
succeed p. 21
sufficient p. 21, p. 10 (insuff.)
sum p. 12 (lump)
summer 5
supply 14 (mov.), p. 21
supplying 29, p. 21
surplus 16 (crop), p. 21
suspend p. 21, p. 14 (order)

t

taels 27
take 29, 30, p. 21
tariff p. 21, p. 7 (duty)
telegrams 1, 2, p. 21
tendency 13
tenders 29, 30, p. 21
terms 20 (opera), 29, 30, 31 (ship), 34, p. 21, p. 15 (pay)
thirds 4 (dates), 26
through p. 21, p. 9 (frt.)
time 3 & 4 (dates), 5 & 6, (Epo), 7 & 8 (hours), 20 (opera), 26 (days), 29 (deliver), 31 (ship), p. 21/2
to-day & morrow 7 (hours)
tonnage 14, p. 22
tons 19, 26/8, 34 (per)
total 20 (opera), 31 (ship), p. 22
towns 32/3
transfer p. 22, p. 19 (rem.), p. 21 (tel.)
transshipment 31, p. 22
trial 31 (ship), p. 22

u

uneasy p. 8 (feeling)
unload 4 (dates)
unsold p. 15 (pos.), p. 19 (sales)
until 4 (dates), 5 (Epo) p. 22
utmost p. 6 (do)

v

valorem p. 22, p. 7 (duty)
value p. 22, p. 8 (face)
via 32/3 (ports)

w

wait p. 22
week 5, 20 (opera) 31 (ship.), p. 22
weekly 31 (ship.)
weight 19, 26, p. 22/3
when 29, 30, p. 23
winter 5
wire 29, 30, p. 4 (commun. interr.), p. 5 (cut), p. 21 (telegr.)
working p, 5 (days), p. 23
worms 15 (crop)

y

yard/s 17, 26/8, 34 (per)
year 5, p. 23
yen 27
yesterday 7 (hours)

Index du texte

les chiffres sans la lettre «p» sont les Nos des tables

a

accepter 1, 17/9, 29/30, p. 24, p. 27 (cond.) p. 42 (propos. & protêt), p. 47 (traite)
accord 1, p. 24
accorder 29/30
achat 17/9, 20/2 (opér.), 23 (±), 29/30, p. 24, p. 43 (quant. ach.)
acheteurs 20/2 (opér.), 23 (±), p. 24, p. 26 (choix), p. 37 (nomin.), p. 41 (prix)
action p. 34 (intenter)
affaire p. 24, p. 48 (urgente)
affirmatir. 1, 7 (heur.)
affrêter 29/30
agir 1, 29/30, p. 24
an/née 5, p. 24
annuler 4, (date), 29/30, p. 29 (date), p. 35 (limite), p. 37/8 (offre), p. 38 (option & ordre)
approvisionn. 14 (mouv.), p. 24
approximat. p. 24, p. 43 (quant.)
argent 13 (mouv.)
arranger 29/30, p. 24
arrivages 14 (mouv.), p. 42 (prix)
arrivée 4 (dates), 7 (heur.). 20, p. 25, p. 29 (date)
assurer 29/30, p. 25
augmenter 29/30 (prix)
aussitôt poss. p. 40
autant poss. p. 25
automne 5
avant 4 (dates), 5 (épo.)

b

baisse 13 (mouv.), 22
balles 18, 26/7, 34 (par)
barils 18, 26/8, 34 (par)
bateau 4 (charger), p. 25, p. 38 (parti)
billet p. 26 (classe)
bon marché p. 35
bord p. 25, p. 34 (jeté), p. 39 (péri)
brevet p. 25, p. 48 (violé)
bruit p. 25, p. 40 (polit.), p. 45 (rum.)
brut p. 25, p. 40, p. 47 (toile)
bushels 26/8

c

caisses 18, 26/7, 34 (par)
campagne 5, p. 26
canal 32 (ports), p. 26
capital 9 & 10, p. 26
centimes 25/6
cents 24/6
pour cent 23 (±), 26, p. 26, p. 40 (plus)
c/m 26
c. & f., *c. f. & a.* 20, 32/4, p. 29
chaleur 15 (sta.) p. 26
chances p. 26, p. 39 (persp.)
change 11 & 13, p. 26
changement 9 (rens.), p. 26
charge p. 26, p. 24 (achet. & vend.)
charger 4 (bat.). p. 26
chemin de fer 34, p. 26, p. 48/9 (vitesse)
chèque 29/30 (payer & rem.), p. 26, p. 39 (payem.)
choléra p. 26, p. 36 (mort)
code p. 27
commandes 1 (corr. & tél.), 4 (dat. exp.), 29/30 (exéc. & annul.), 31 (exp.), p. 27, p. 35 (lettre & livraison), p. 38 (ordres)
compétence 9 & 10
comptant 34, p. 27
compte p. 27, p. 26 (chèque)
conclure 23 (±), 29/30, p. 35 (limites)
concurrence 20, 23 (±), p. 27, p. 38 (offre), p. 41 (prix)
conditions 20, 31 (exp.), 34, p. 27/8, p. 27 (compt.), p. 39 (payem.)
confirmer 1 (corr.), p. 28
connaissements p. 28
consentement p. 28, p. 24 (accord)
consignation 29/31 (exp.), p. 28
consommation 14 (mouv.), p. 28
contrat 29/30 (renouv.), p. 28, 41/2 (prix)
convention p. 28, p. 41 (prix)
correspond. 1, p. 28, p. 35 (lettres)
cotes 13 (mouv.), p. 29
coter 29/30
courrier p. 29, p. 35 (malle)
coût p. 29, p. 41/2 (prix)
c. & f., *c. f. & a.* 20, 32/4, p. 29
couvrir, 20/1, 29/30, p. 29
crédit 9 & 10, 11 (lettre), 29/30 (accord.), p. 29
cwt. 19, 26, 34 (par)
cyclone 15 (stat.)

d

d. 23 (±), 24, 26
dates 1, 3, 4, p. 25 (arr.), p. 29
débarquement 4 (dates), p. 40 (port.)
débordé p. 27 (comm.), p. 38 (ordres)
décharger 4 (dat.)
déclaration p. 29/30, p. 48 (valeur)
découvert 9 & 10, 29 & 30 (vendre), p. 30, p. 29 (créd.), p. 42 (prix)
déficit p. 30, p. 40 (poids)
dégâts p. 30
délai 5 & 6 (épo.), 17/20, 31 (exp.), p. 30, p. 35 (livr.), p. 41 (prix), p. 47 (terme)
délivrer 29/30, p. 44 (remb.)
demain 7
demande 14 (mouv.), 21/2, p. 30, p. 42 (prix)
départ 4 (date), 7 (heur.), p. 30, p. 29 (date), p. 38 (partir)
déposer 29/30, p. 30
désirer 29, p. 35 (man.)
dès 5 (épo.)
détail p. 30, p. 35 (lettre)
disposition p. 30, p. 45 (retard)
documents p. 30, p. 39 (payem.)
dollars 12, 25, 27/8
domicile 34, p. 31
dommage 15 & 21 (récolte), p. 31, p. 30 (dégâts)
douane p. 31, p. 30 (dédouan.)
douzaine 26, 34 (par)
droits 34, p. 31
dû 3, p. 31
ducroire p. 31, p. 27 (comm.)

e

écart 23 (±), p. 31
échantillons 16 (stat. qual.), 20 & 31 (qual.), p. 31, p. 43 (qual.)
échéance 3 (dû), p. 31
effets 11/2, 29/30 (accept, payer & tirer), p. 31, p. 39 (payem.), p. 42 (prot.), p. 47 (traites)
emballage 34, p. 31
embarquem. 14 (mouv.), 20/2 & 31 (exp.), p. 31, p. 39 (permis), p. 40 (port.)
emmagasiner p. 31, p. 35 (mag.)
encaisser 29/30, p. 31
enchères 20, p. 31/2
engager 29/30
environ 4 (date), 5 (épo.)
époques 5 & 6, 20 (opér.), 31 (exp.), p. 32

équipage p. 32, p. 39 (péri.)
erreur 1 & 2 (corr. & tél.), p. 32
escale p. 32, p. 40 (port.)
escompte 13 (mouv.), 34, p. 32, p. 27 (compt.)
escompter 29/30
établi 9 & 10, p. 32
état p. 32, p. 45 (santé)
étendue 29/30, p. 32
ex p. 48 (vap.)
exécuter ordre 29/30, p. 38
existences 14 (mouv.)
expéditions 1, 3 & 4 (commande), 5 & 6 (épo), 14 (mouv.), 16 (stat.), 17/22 (opér.), 29/33, p. 32, p. 27 (col. post.), p. 31 (emb.), p. 44 (remb.), 48/9 (vit.), 49 (wag.)

f

fabrique 34 (usine), p. 33
facture 3 (date), 11 (mont.), 30 (remis), p. 33, p. 28 (copie & connaiss.)
faculté 23 & 32/3 (opt.), p. 33, p. 26 (choix), p. 38 (opt.)
faillite 9, p. 33, p. 46 (susp.)
faire 29/30
faut 29 (nécess.), p. 33
faveur p. 33, p. 41/2 (prix)
fièvre p. 36 (mort.)
finance 9/12
florins 26/8
flottant 14 (mouv.), p. 33, p. 42 (prix)
forfait 31, p. 33
fluctuat. 13 (mouv.), p. 26 (change)
fournir (livrer) 29 & p. 35
fournir (tirer) voir traites
fractions 4 (date), 23 (±), 26
frais p. 33, p. 24 (ach.), p. 30 (déduct.)
franco 20 & 32/4 p. 40 (port)
francs 12, 25, 27/8
frêt 13 (mouv.), 14 (tonn.), 29/30 (engag.), 31 (exp.), 32/4, p. 33, p. 42 (prix)
froid 15 (stat.), p. 33
frontière p. 33, p. 30 (dédouan.)
fûts 18, 26/8, 34 (par), p. 33

g

gallons 26/8, 34 (par)
garantie 29/30, 31 (qual.), p. 33, p. 39 (payem.), p. 42 (pur), p. 43 (qual.)
gare 34, p. 33
gelée 15, p. 33
grosses 26, 34 (par)

h

hausse 13 (mouv.), 21, 23 (±)
heures 7 & 8
hier 7
hiver 5

i

in. 26
inconnu p. 28 (conn.)
insectes 15 (stat.)
intenter p. 34
intention 29, p. 34
intervenir 29

j

jour 26, p. 34
jusqu'à 4 (dat.), 5 (épo.), p. 34

k

kilo 19, 26/8, 34 (par)

l

légalisat. p. 34, p. 26 (cert.), p. 33 (fact.)
leste 4
lettres 1, 3, 11 & 34 (crédit), p. 35
limiter 29/30
limites 20/2, 23 (±), p. 35
Lire it 27
liste p. 35, p. 42 (pr. cour.)
livraison 4 (date), 5 & 6 (épo.), 17/9 (délai), 20, 29/30 (délivrer, fourn. & prendre), p. 35, p. 30 (docum.), p. 44 (remb.), p. 45 (retard)
£ stg. 12, 24, 26/8
lbs 19, 26, 34 (par)
locustes 15 (réc.)
lot 6 (épo.), p. 35, p. 38/9 (parties)

m

malle p. 35, p. 29 (courr.)
marcs 12, 25, 27/8
marché 13 (mouv.), 23 (±), p. 35/6
marge 23 (±), p. 36
marque p. 36
matière prem. 13 (mouv.), p. 36
matin 7 & 8, p. 36
maximum 17/9, p. 40 (poids), p. 43 (quant.)
mètres 17, 26, 34 (par)
mieux 1 (faire), p. 36
m/m 26
millions 28
milreis 27
minimum 17/9 (quant.), p. 40 (poids), p. 43 (quant.)
mode 34, p. 36, p. 39 (payem.)
moins 23 (±), p. 36, p. 40 (plus)
mois 5 & 6, 20, 31 (exp.), p. 36
moitié 4 (dates), 11 (fin.), 20 (opéra), 31 (exp.), p. 36
mots 2, p. 36
moyens 9 & 10 p. 26 (capital)
moyenne 16 (stat.), p. 36, p. 43 (quant.)

n

naufrage p. 37, p. 39 (péri)
nécessaire 29, p. 37
négativ. 1 (corr.), 7 (heur.)
négociations p. 37, p. 48 (ventes)
net 34, p. 37, p. 40 (poids)
nolis (frêt) 13 (mouv.), 14 (tonn.), 29/30 (engag.), 31/4, p. 33
nom p. 37, p. 25 (bat.), p. 48 (vap.)
nominal 13 (mouv.), p. 37, p. 41 (prix)
non 1 (corr.), 7 (heur.), p. 37
nouvelles p. 37, p. 43 (rapp.)
noyé p. 37, p. 39 (péri)

o

obtenir 29/30, p. 37
occasion 31 (frêt), p. 37, p. 24 (achet.), p. 33 (frêt), p. 48 (ventes)
offres 1 & 2, 4 (dates), 6 (épo), 7 (heur.), 14 (mouv.), 17/22, 23 (±), 29/30 (accept. & offrir), p. 37/8, p. 41 (prix)
option 23 (±), 32/3 (ports), p. 38
ordres 1 & 4 (exp.), 30 (comm.), 29/30 (exéc.), 31, p. 38, p. 27 (comm.), p. 28 (contre-o.), p. 35 (livr.)
oui 1, 7, p. 38

p

partir 4 (date), 7 (heur.), p. 38, p. 30 (dép.)
partie 6 (épo), p. 38/9, p. 35 (lots)
payements 3 (dû), 10, 12, 29/30, 34 (cond.), p. 39, p. 26 (chèque), p. 34 (intér.), p. 40 (poursuite), p. 46 (susp.), p. 47 (traites)
péri p. 39, p. 37, (naufr. & noyé)
perspectives p. 39, p. 26 (chances)
perte 9, p. 39, p. 34 (intérêts)
pfg. 25/6
piastres 27
pieds angl. 26, 34 (par)
sur place 20, p. 40
pluies 15 (stat.) p. 40
poids 19, 26/8, p. 40
police p. 25 (ass.)
port/s 32/3, 34, p. 40
poste p. 40, p. 24 (adr. p. rest.)
pour cent 23 (±), 26, p. 26, p. 40 (plus)
pouvoir 29, p. 41
prendre 29/30
pressé p. 41, p. 47/8 (urg.)

prêt 4 (*dates*), 29, p. 41
printemps 5
prix 13 (*mouv.*), 17/22, 23 (±), 34 (*s'entend*), p. 41/2, p. 24 (*achats*), p. 37 (*occas.*), p. 41 (*préfér.*), p. 48 (*ventes*)
procès p. 42, p. 34 (*intent.*)
production 14 (*mouv.*), 16 (*stat.*), p. 42 (*prix & prod.*)
profits & pertes p. 27 (*compte*)
prolonger 4 (*date*), 29/30
proposer 1, 29/30, p. 42
prorogation 4 (*date*), 29/30 (*prol.*), p. 42
protester 29/30, p. 42, p. 39 (*payem.*)

q

quai 34
qualité 13 (*prix*), 14 (*dem.*), 15/6 (*stat.*), 20, 29 (*garantir*), 31 (*exp.*), p. 42/3, p. 31 (*échant.*), p. 41 (*prix*), p. 45 (*retours*)
quand 29/30, p. 43
quantité 17/20, 29/30 (*étendue*), 31, p. 43, p. 35 (*livrée*), p. 47 (*total*)
quarts 4 (*dates*), 26
qrs 26/7, 34 (*par*)

r

rachats p. 43, p. 42 (*prix*)
rapports p. 43, p. 37 (*nouvelles*)
réalisations 22, p. 43, p. 42 (*prix*)
recettes 14 (*mouv.*), 21/2
récolte 15/6, 20/2, p. 43/4, p. 42 (*prix*)
recommand. 1 & 29, p. 44, p. 35 (*lettre*)
rectifier 1 & 2, p. 32 (*erreur*)
reçu 1 (*corr.*), 12 (*mont.*), 30, p. 43/4, p. 32 (*envoi*), p. 35 (*march.*)
réduction 29/30 (*prix*), p. 44, p. 43 (*rabais*)
refus p. 44, p. 42 (*qual.*), p. 31 (*éch.*), p. 45 (*retours*)
régler 29/30
règlement 34, p. 44
remboursement p. 44, p. 39 (*payem.*)
remettre 11, 12, 29/30, p. 44, p. 31 (*encaiss.*), p. 35 (*mand. p.*), p. 45 (*solde*)
remplacer 29/30, p. 44
rendement 15/6 (*stat.*), p. 44
renouveler 1, 7 & 23 (*offre*), 29/30, p. 37/8 (*offre*)
renseignements 9 & 10, 29/30 (*télégr.*), p. 44, p. 24 (*s'abst.*), p. 29 (*crédit*)
répéter 2 (*tél.*), 29/30 (*transact.*)
répondre 1, 2, 4 (*offres*), 7 (*heur.*), p. 44, p. 46 (*télégr.*)
reste 4, 20, 31, p. 44, p. 46 (*suivra*)
résultat p. 44/5
retard p. 45, p. 34 (*intérêts*)
retourner p. 45, p. 42 (*qual.*)
revient p. 42 (*prix*)
révocation p. 45, p. 37 (*offres*), p. 48 (*val.*)
roubles 27
roupies 27
rumeurs 9, p. 45, p. 25 (*bruits*)

s

sacs 18, 26, 34 (*par*)
saison 5, 16, p. 26 (*camp.*)
satisfaire p. 45, p. 27 (*client*)
sauvé p. 45, p. 32 (*équip.*)
semaines 5, 20, 31 (*exp.*), p. 45
semestre p. 45
sh. 23 (±), 26
situation 9, 10, p. 45, p. 38 (*ordres*), p. 44 (*renseign.*)
soins p. 24 (*adr.*)
soir 7, 8, p. 45
solde 4 (*dates*), 11 (*fin*), 20 & 31 (*reste*), p. 45, 27 (*compte*), p. 39 (*payer*)
solvabilité 9, 10, p. 46, p. 29 (*crédit*)
soumission 29/30, p. 46
station 34, p. 46
stipuler 30, p. 27 (*cond.*), p. 32 (*étendue*), p. 35 (*manière*)
stocks 14 (*mouv.*), 21/2, p. 42 (*prix*)
substituer 29/30, p. 46
sur 11
suspendre payem. 9, p. 46, p. 33 (*faill.*), p. 38 (*ordres*)

t

taels 27
tard 4 (*dates*), 5 (*épo.*), p. 46
tarif p. 46, p. 31 (*droits*), p. 42 (*pr.-cour.*)
télégraphier 29/30, p. 46, p. 36 (*mots*)
télégrammes 1, 2, p. 46
temps 3/8, p. 46/7
tendance 13 (*mouv.*)
tiers 4 (*dates*), 11 (*fin.*), 26
tirer 29/30 (*voir traites*)
tonnage disp. 14
tonnes 19, 26/8, 34 (*par*)
traites 9/12, 29/30 (*accept., payer, rem., tirer*), p. 47, p. 39 (*payem.*), p. 42 (*prot.*)
transborder 31, p. 47
trimestre 5

u

urgent p. 47/8, p. 41 (*pressé*)
usine 34, p. 48, p. 33 (*fabr.*)

v

valable & vigu. p. 48
valeur p. 48, p. 25 (*assur.*), p. 30 (*déclar.*)
vapeur 4 (*dates*), 20, 31/3, p. 48, p. 38 (*partir*)
vendeurs 20/2, 23 (±), p. 24 (*achet.*), 41 (*prix*)
ventes 17/22, 23 (±), 29/30, p. 48, p. 32 (*enchères*), p. 42 (*prix*), p. 43 (*quant.*)
vermoulure 15 (*stat.*), p. 48
via 31/3
vigueur & val. p. 48
voilier 31/3, p. 49
vue 11, p. 49

w

wagon 34, p. 49

y

yards 17, 26/8, 34 (*par*)
yen 27

Mutilation Key

for "Vollers' A.-F. Code"

Clef des Mutilations

du Code Télégraphique:

« Vollers' A.-F. Code »

compiled and published by

compilé et édité par

R. VOLLERS

Printed by / *Imprimé par* } E. Mühlthaler's Buch- u. Kunstdruckerei A.G., Munich

1904

The particular system on which my Code is based renders it extremely easy to trace mutilations by means of this key.

It is nearly impossible that incorrect transmission or mutilations of my code ciphers could have serious consequences as

I All my ciphers differ from each other by at least 2 letters.

II Dangerous resemblances with respect to the "Morse" symbols **do not exist in my cyphers.**

For instance:

in case I I have avoided resemblances such as:

tirar (table 32) and **virar**
tiraria „ „ „ **tirarian**

(the Spanish termination "arian" also occurs in my Code viz: in tables 2 and 27 under group 28, but in conjunction with quite another group of Spanish roots than "aria" which latter belongs to group 23 = tables 17, 20/2, 28—33 and 36).

in case II I have avoided resemblances such as:

bordastila (table 13) and bordavila as "st" (• • • | —) and "v" (• • • —) might very easily clash in telegraphic transmission.

(this Italian termination "avila" also exists in my Code but in connection with quite a different group of Italian roots viz: group 21 — table 26, whereas "astila" belongs to group 14 = tables 5, 13, 23 and 25).

Example for the use of this key: the word "apestada" reaches as "agestada": look up termination "ada" under "a" page 4 of this key; the group 23 to which you are referred indicates that the root exists in one of the tables 17, 20/2, 28—33 or 36; among the roots under "t" page 18 of this key you will find then at a glance that the word could not be any other than "apestada".

If a mutilation occurs in the termination the right termination will easily be traced among the others of the same table.

Le système spécial d'après lequel mon code a été compilé rend très facile de trouver la trace des mutilations au moyen de cette clef.

Il est presque impossible qu'une **transmission incorrecte** ou une **mutilation** des chiffres de mon code puisse avoir de graves **conséquences car**:

I mes chiffres diffèrent tous entre eux par 2 caractères au-moins.

II Les ressemblances dangereuses en ce qui concerne les symboles "Morse" n'existent pas dans mes chiffres; par exemple:

dans le cas I J'ai évité des ressemblances comme celles-ci: **tirar** (table 32) et **virar**
tiraria „ „ „ **tirarian**

(la terminaison espagnole «arian» existe aussi dans mon Code savoir: aux tables 2 et 27 sous groupe 28; mais en connection avec tout un autre groupe de radicaux espagnols que «aria», ce dernier appartenant au groupe 23 = tables 17, 20/2, 28—33 et 36.)

dans le cas II J'ai évité des ressemblances de symboles "Morse" comme celles-ci:
"borda**st**ila" (table 13) et "borda**v**ila" vu que "st" (• • • | ——) et "v" (• • • ——) pourraient facilement se confondre pendant la transmission télégraphique.

(cette terminaison italienne «arila» se trouve aussi dans môn code mais en rapport avec un groupe tout différent de radicaux savoir: groupe 21 = table 26, tandis que «astila» appartient au groupe 14 = tables 5, 13, 23 et 25.)

Exemple pour l'emploi de cette clef.

Le mot "**apestada**" arrive comme "**agestada**":
cherchez la terminaison "**ada**" sous "**a**" page 4 de cette clef; le groupe 23 auquel on vous réfère alors, indique que le radical existe à une des tables 17, **20/2, 28—33** ou 36; vous trouverez en un clin d'oeil parmi les radicaux sous "t" page 18 de cette clef que le mot ne pourrait être autre que "apestada".

Si la mutilation arrive dans la terminaison, la terminaison correcte sera trouvée facilement parmi les autres de la même table.

"Morse" Symbols. *Symboles «Morse.»*

a	• ——	i	• •	r	• —— •
b	—— • • •	j	• —— —— ——	s	• • •
c	—— • —— •	k	—— • ——	t	——
d	—— • •	l	• —— • •	u	• • ——
e	•	m	—— ——	v	• • • ——
é	• • —— • •	n	—— •	w	• —— ——
f	• • —— •	o	—— —— ——	x	—— • • ——
g	—— —— •	p	• —— —— •	y	—— • —— ——
h	• • • •	q	—— —— • ——	z	—— —— • •

The following instances show how mutilations would occur through omission of the | (separation bar between the letters) or through a mistake in its application

Les exemples suivants expliquent comment les mutilations arriveraient par suite de l'omission de | (barre de séparation entre les lettres) ou par une erreur dans son application.

• —	**a**	= et
		ae = r
		ai = l
		am = j
		an = p = ate
		at = w
— • • •	**b**	= de = ni = ts
— • — •	**c**	= nn = tr
— • •	**d**	= ne = ti
		de = b
		dt = x
•	**e**	ea = u
		ee = i
		ed = l
		ei = s
		em = w
		en = r
		er = in = f
		es = ii
		eu = v
• • — • •	**é**	= el = id
		ui = fe
• • — •	**f**	= er = eae = in = ue
— — •	**g**	= me = tn
		ge = z
		gt = q
• • • •	**h**	= ii = es = se
• •	**i**	= ee
		ia = v
		ie = s
		in = f = er
		it = u
• — — —	**j**	= eo = am = att
— • —	**k**	= nt = ta = tet
• — • •	**l**	= ai = ed = re
— —	**m**	= tt
		ma = q
		me = tn = g
		mi = td = mee = ge
		mm = ot = to
		mt = o
— •	**n**	= te
		na = x
		ne = d
		ni = b
		nm = y
		nn = c
		nt = k
— — —	**o**	= tm = mt
• — — •	**p**	= an = eg = ette = eme = we
— — • —	**q**	= ma = gt = met
• — •	**r**	= ae = en
		re = l
• • •	**s**	= ei = ie
		se = h
		st = v
—	**t**	ta = k
		td = z
		te = n
		ti = d
		tn = g
		tr = c
		ts = b
		tt = m
		tu = x
• • —	**u**	= it = ea
		ue = f
		ui = é
• • • —	**v**	= eu = ia = st
• — —	**w**	= at = em
— • • —	**x**	= na = tu
— • — —	**y**	= nm = kt
— — • •	**z**	= td = mi = mee = ge

Terminations classed by the last letter

the terminations have not more than 6 letters

a

	group:
-aba	28
-abala	28
-abanla	22
-acela	21
-ada	23
-adla	28
-adola	23
-ala	21
-ammola	21
-andola	28
-anla	23
-arala	23
-aranla	28
-arela	28
-aria	23
-ariala	28
-arla	11
-armela	21
-aronla	22
-astela	22
-astila	14
-atala	21
-atela	11
-ava	11
-avala	12
-avanla	21
-avila	21
-ela	26
-enla	28
-era	11
-iamla	11
-iamola	21
-icela	16
-ila	11
-imela	21
-ircela	29
-irla	29
-ironla	29
-istila	29
-itala	29
-iva	29
-ivanla	29
-ivola	29
-ola	28
-olla	12
-otela	24
-ovela	17

d

	group:
-ad	28

e

	group:
-abame	28
-abanme	25
-abare	1
-abase	23
-abaste	28
-acene	11
-adale	23
-adme	23
-adome	28
-adote	23
-aile	11
-ale	28
-amele	24
-ammone	14
-andae	2
-andole	14
-andome	22
-andone	21
-andote	28
-anle	21
-anme	28
-anse	23
-ansene	14
-ante	11
-antene	21
-arame	23
-aranme	28
-arante	23
-arate	28
-are	11
-areme	28
-arere	1
-arete	23
-ariame	28
-ariase	22
-arle	23
-armene	14
-arne	21
-aronle	14
-aronse	22
-aronte	28
-arse	28
-arsene	21
-asele	19
-asse	11
-aste	23
-astele	21
-asteme	22
-astene	14
-astine	21
-atae	5
-atale	11
-atane	21
-ate	21
-atile	21
-atone	11
-atote	1
-aturae	4
-avale	21
-avate	11
-avine	11
-avisse	4
-avole	11
-avone	21
-ebare	7
-endae	7
-endone	29
-enme	26
-ense	28
-ente	29
-erebbe	14
-erere	7
-ereste	21
-erete	11
-ese	26
-ete	28
-iamole	29
-iamone	14
-iate	11
-icene	21
-imene	11
-immone	29
-iranle	29
-irebbe	29
-irete	29
-irmene	29
-irne	29
-isse	29
-istene	29
-ite	29
-itele	29
-itone	29
-itote	8
-ivale	29
-ivine	29
-ole	11
-ome	23
-omele	16
-omene	21
-one	21
-onne	20
-ose	28
-osele	24
-otele	21
-ovene	11

i

	group:
-aivi	11
-ali	11
-amini	1
-ammoci	21
-ammovi	14
-andi	1
-andoci	14
-andomi	21
-ansi	11
-antesi	14
-anti	21
-arci	11
-ari	1
-armi	21
-aronsi	14
-aronvi	21

Terminaisons classées par la dernière lettre

les terminaisons n'ont pas plus de 6 lettres.

	group:
–assi	21
–astemi	21
–astevi	14
–asti	11
–astici	14
–atami	21
–atasi	11
–ateci	21
–atemi	11
–atili	11
–atisi	21
–atoti	21
–atovi	11
–aturi	5
–avaci	11
–avanci	21
–avasi	21
–aveli	16
–avi	21
–avici	21
–avimi	11
–avisti	4
–avomi	21
–avoti	11
–emini	7
–endi	7
–endovi	29
–entesi	29
–erai	21
–erassi	21
–erei	11
–eresti	14
–iamoci	14
–iamoti	21
–iamovi	29
–iamvi	21
–icci	29
–iceli	21
–ici	21
–imi	11
–immoti	29
–irai	29
–iransi	29
–iresti	29
–irommi	29
–irossi	29
–irotti	29
–isti	29
–istici	29
–itami	29
–iteci	29
–iteli	19
–itivi	29
–itosi	29
–ivaci	29
–ivanti	29
–ivimi	29
–ivoti	29
–ivvi	29
–occi	21
–oceli	11
–omi	21
–ommi	18
–ossi	11
–otti	13
–oveli	21
–ovi	11

m

–abam	1
–am	7
–andam	6
–andum	2
–antem	1
–antium	6
–atarum	4
–atum	5
–averam	4
–endum	10
–entem	7
–entium	9
–erem	7

n

–aban	23
–aren	28
–arian	28
–aron	23
–asen	23
–en	26

o

–abalo	23
–abanlo	28
–abo	1
–acelo	16
–adlo	23
–adolo	28
–amelo	21
–ammo	11
–ammolo	14
–ando	15
–andolo	22
–anlo	28
–ano	21
–anto	1
–aralo	28
–aranlo	22
–arelo	23
–arialo	22
–arlo	21
–aronlo	28
–arono	11
–assero	14
–assimo	14
–astelo	28
–astilo	21
–atelo	21
–ato	11
–atolo	11
–avamo	11
–avanlo	14
–avano	21
–avero	5
–avilo	11
–avolo	21
–elo	28
–endo	29
–endolo	29
–enlo	26
–eranlo	21
–eranno	14
–eremo	11
–eremmo	21
–ero	21
–eselo	24
–iamlo	21
–iamo	11
–iamolo	14
–illo	29
–ilo	21
–imelo	16
–immo	29
–immolo	29
–ino	11
–iranno	29
–iremo	29
–irmelo	29
–iro	29
–irollo	29
–irono	29
–issero	29
–istelo	29
–ito	10
–itolo	29
–ivano	29
–ivilo	29
–ocelo	21
–ollo	21
–olo	23
–omelo	24
–otelo	13
–unto	8

r

–abatur	6
–abimur	6
–amur	1
–andior	6
–antur	1
–ar	23
–aretur	6
–arier	1

Terminations classed by the last letter

Terminaisons classées par la dernière lettre

	group:
-atior	5
-atur	7
-ebatur	9
-endior	9
-entur	7
-eretur	9
-etur	1
-imur	7
-or	3

s

-abais	23
-abalas	22
-abales	28
-abamos	28
-abamus	6
-abanos	22
-abaris	6
-abis	1
-abitis	6
-adas	28
-adlas	23
-adles	28
-adnos	28
-adolas	28

	group:
-adoles	22
-adonos	22
-ados	23
-ais	23
-ales	23
-amonos	28
-amos	28
-amus	7
-andius	6
-andos	1
-anlas	28
-anlos	23
-ans	1
-antis	1
-arais	28
-aralas	28
-arales	22
-aramos	22
-aranos	28
-aras	23
-areis	23
-arelas	22
-areles	28
-aremos	28
-aremus	6
-ares	1

	group:
-aretis	6
-ariais	22
-arieis	28
-arlos	28
-arnos	23
-aseis	28
-asemos	22
-asteis	22
-atis	7
-atius	5
-atos	5
-aturos	4
-avimus	4
-ebamus	9
-ebas	7
-ebatis	9
-elas	28
-elos	26
-emonos	27
-emus	1
-endius	9
-endos	7
-enlas	26
-enles	28
-ennos	26
-enos	28

	group:
-ens	7
-entis	7
-eremus	9
-eretis	9
-eris	7
-es	1
-etis	1
-is	10
-olas	23
-oles	28
-omelos	28
-onos	23

t

-abant	1
-ant	7
-arent	1
-at	1
-averit	4
-avit	5
-ebant	7
-ent	1
-erent	7
-et	7

r—t

Groups 1—6 (Latin)

Group 1:	refer to / *voir*	p. 1—49, tables 1, 4, 7, 9, 12, 15, 24, 35
„ 2:	„	tables 7, 9, 12, 24, 35
„ 3:	„	p. 1—49, tables 1, 4, 9, 12, 15, 35
„ 4:	„	p. 1—49, table 1
„ 5:	„	p. 1—49, tables 1, 9, 12, 35
„ 6:	„	p. 1—49, tables 1, 4, 15

see the following roots alphabetically arranged by their last letter

voir les radicaux suivants rangés par ordre alphabétique de leur dernière lettre

Roots of groups 1—6 *Radicaux des groupes 1—6*

b-

allab-
baub-
dealb-
exalb-
glob-
inalb-
incib-
prob-
seb-

c-

abdic-
abloc-
albic-
amic-
avoc-
cac-
deocc-
desec-
educ-
effoc-
eloc-
emanc-
emusc-
exsec-
fodic-
foed-
fusc-
indic-
inesc-
inocc-
inunc-
invoc-
judic-
loric-
nec-
obunc-
offoc-
opac-
oppic-
ovic-
plic-
pumic-
rauc-
runc-
sacc-
sevoc-
spic-
spurc-
sulc-
tunic-
visc-

d-

abnod-
desud-
emend-
enud-
exsud-
exund-
fluid-
fraud-
innod-
lapid-
laud-
livid-
madid-
nitid-
nod-
pand-
pavid-
ped-
praed-
rigid-
solid-
sucid-
sud-
tard-
tepid-

e-

abige-
balte-
calce-
clipe-
comme-
cune-
gale-
line-
reme-

f-

irruf-

g-

ableg-
adjug-
allig-
arrog-
aurig-
calig-
delig-
deneg-
derog-
dijug-
effug-
elong-
erug-
evag-
evulg-
greg-
indag-
irrig-
jug-
levig-
lig-
mitig-
plag-
purg-
refug-
relig-
rog-
sejug-
volg-

h-

lymph-

i-

basi-
brevi-
capri-
cruci-
ebri-
elogi-
furi-
glaci-
impi-
infri-
inhi-
jurgi-
lani-
longi-
maci-
mini-
nunti-
pari-
radi-
sati-
sauci-
soci-
somni-
spoli-
tali-
terti-
viti-

l-

adhal-
advel-
aemul-
affl-
ambul-
amyl-
ancl-
angul-

Roots of groups 1—6 *Radicaux des groupes 1—6*

anhel–
assol–
bell–
cael–
capul–
confl–
degul–
depil–
desol–
devel–
diffl–
dupl–
edol–
egel–
exhal–
expil–
exsol–
exul–
famul–
figul–
focil–
gel–
hal–
immol–
incal–
infl–
insol–
jubil–
obvel–
ocul–
perfl–
popul–
profl–
recol–
refl–
regel–
rutil–
sibil–
simpl–
stell–
suffl–
sugil–
tabul–
tripl–
tumul–
ulul–
ustil–
vall–
vigil–
viol–

m-

adam–
anim–
arm–
autum–
crem–
deam–
dedom–
flamm–
form–
hiem–
hum–
illim–
infam–
inhum–
intim–
negum–
obhum–
opim–
plasm–
plum–
racem–
redam–
sim–
spum–
stigm–
summ–

n-

adun–
alien–
amoen–
ann–
atton–
carin–
clin–
coen–
coron–
damn–
dapin–
dign–
don–
fren–
inorn–
inton–
lacun–
lumin–
man–
omin–
ordin–
orn–
redon–
regn–
reorn–
resan–
sagin–
seren–
stagn–
supin–
tamin–
tran–
urin–
venen–

p-

alap–
aucup–
crisp–
culp–
obsip–
occup–
palp–
pomp–
rusp–
sirp–
stip–
supp–
torp–
usurp–

r-

accur–
adaer–
ador–
aer–
agger–
amar–
appar–
aspir–
augur–
camer–
castr–
celer–
clar–
coir–
color–
cribr–
deaur–
devor–
dur–
ejer–
emigr–
err–
evir–
exjur–
exor–
flagr–
funer–
gener–
gyr–
hilar–
ignor–
imper–
incer–
indur–
inerr–
infor–
intr–
iter–
labor–
litur–
lucr–
lustr–
macer–
matur–
memor–
minor–
mir–
moder–
narr–
nigr–
numer–
obar–
obdur–
oner–
patr–
pejor–
plor–
quadr–
reser–
rigor–
separ–
sopor–
sper–
stupr–
subar–
temer–
tetr–
toler–
uber–
ulcer–
umbr–
vapor–
vibr–
vor–

s-

camps–
dens–
evers–
excus–
exoss–
fals–
foss–
gyps–
lass–
mers–
muss–
obes–
pis–
press–
puls–
recus–
spiss–
spons–
trus–

t-

accit–
adapt–
agit–
alt–

Roots of groups 1—6 *Radicaux des groupes 1—6*

amput–
annot–
argut–
ariet–
audit–
cant–
cert–
cit–
coact–
coept–
comit–
crust–
cunct–
dat–
demet–
denot–
deopt–
deput–
devit–
dict–
dilat–
edent–
effet–
elect–

emut–
enat–
epot–
equit–
excit–
exopt–
exput–
facet–
ferit–
flut–
fugit–
gest–
habit–
hebet–
imit–
immut–
invit–
jact–
just–
laet–
lent–
maest–
marit–
mot–

mult–
olet–
onust–
opert–
opput–
opt–
oscit–
plant–
port–
put–
rapt–
recit–
ret–
risit–
ruct–
salut–
scrut–
spect–
tempt–
tract–
umect–
vast–
vect–
veget–

velit–
vocit–

u-

aequ–
aestu–
artu–
cernu–
febru–
idu–
iniqu–
liqu–
sinu–
tenu–
vacu–
vetu–

v-

adlev–
curv–
delav–
deriv–

elev–
enerv–
excav–
grav–
ignav–
larv–
nav–
priv–
salv–
serv–

x-

anax–
crax–
divex–
elix–
flex–
lux–
malax–
relax–
tax–
vex–

Groups 7—10 (Latin)

Group 7:	refer to / *voir*	tables 3, 6, 11	see the following roots alphabetically arranged by their last letter *voir les radicaux suivants rangés par ordre alphabétique de leur dernière lettre*
„ 8:	„	table 11	
„ 9:	„	„ 3	
„ 10:	„	tables 6, 11	

Roots of groups 7—10 *Radicaux des groupes 7—10*

b-

adbib–
bib–
glub–
imbib–
lamb–
nub–

obbib–
obnub–
scab–
scrib–

c-

adduc–
disc–

edic–
induc–
nasc–
obduc–
parc–
posc–
scisc–
seduc–

d-

accud–
aded–
adind–
adrad–
allid–
ambed–

arrod–
caed–
clod–
coadd–
comed–
decid–
derad–
divid–

Roots of groups 7–10 *Radicaux des groupes 7–10*

elud-
erod-
excud-
exed-
exlid-
find-
illud-
incud-
invad-
irrad-
obed-
oblid-
obrod-
occid-
pend-
pered-
redd-
relid-
scind-
spand-

g-

adig-
ambig-
arrig-
attig-
cing-
desug-
deteg-
deung-
eling-
emerg-
emung-
erig-
exsug-
flig-
frang-
fung-
inig-
integ-
inung-
leg-
obteg-
offig-
pang-
perag-
reag-
recog-
refig-
releg-
sparg-
subig-

h-

aveh-
inveh-
reveh-

l-

avell-
cill-
coal-
emol-
incol-
occul-
pell-

m-

abem-
absum-
coem-
dirim-
exim-
frum-
insum-
prem-
resum-
revom-

n-

accin-
decan-
degun-
delin-
depon-
expon-
gign-
illin-
impon-
oppon-
pon-
stern-

p-

clep-
erep-
insip-
irrep-
rump-
sculp-
serp-
strep-

r-

adur-
ambur-
asser-
atter-
averr-
eger-
exser-
exur-
inger-
inur-
obger-
perur-
quaer-
subur-

s-

deps-
juss-
pins-

t-

emet-
exist-
expet-
flect-
impet-
mitt-
nect-
oppet-
pet-
repet-
sist-

u-

acu-
allu-
argu-
battu-
consu-
coqu-
corru-
deflu-
deru-
dilu-
efflu-
futu-
imbu-
intu-
irru-
metu-
obsu-
proru-
renu-
sequ-
sublu- 7–10

v-

depuv-
evolv-
solv-

x-

retex-

Groups 11—20 (Italian)

Group **11**:	refer to / *voir*	tables 5, 8, 10, 13, 14, 16, 18, 19, 23, 25	see the following roots alphabetically arranged by their last letter *voir les radicaux suivants rangés par ordre alphabétique de leur dernière lettre*
„ **12**:	„	„ 5, 8, 10, 13, 16, 18, 19, 23, 25	
„ **13**:	„	„ 5, 8, 10, 14, 16, 18, 19, 23, 25	
„ **14**:	„	„ 5, 13, 23, 25	
„ **15**:	„	„ 5, 10, 13, 14, 16, 18, 19, 23, 25	
„ **16**:	„	„ 5, 8, 10, 13, 14, 18, 19, 23, 25	
„ **17**:	„	„ 5, 8, 10, 16, 18, 19, 23	
„ **18**:	„	„ 5, 8, 10, 14, 23	
„ **19**:	„	„ 8, 10, 14, 18, 19	
„ **20**:	„	„ 5, 10, 23	

Roots of groups 11—20 *Radicaux des groupes 11—20*

b-

acerb–
adorb–
arrub–
delib–
dirob–
fromb–
gabb–
giamb–
nerb–
piomb–
rob–
sturb–
zomb–

d-

abond–
accod–
affid–
ammod–
arred–
bend–
biad–
biond–
blind–
bord–
chiod–
comod–
covid–
denod–
dinud–
frod–
grid–
grond–
guad–
inond–
lard–
lod–
lucid–
mand–
nud–
ossid–
parod–
pond–
pred–
rilod–
sband–
scud–
snod–
stend–

f-

biff–
ceff–
ciuff–
insof–
intuf–
raff–
scaf–
solf–
tuff–

j-

abbuj–
ammaj–
appaj–
dimoj–
inaj–
incoj–
soj–
spaj–

l-

accul–
adul–
aggel–
anell–
arrol–
asol–
attel–
avall–
avvil–
bajul–
boll–
burl–
cull–
desal–
difil–
dipel–
duell–
esil–
fucil–
giel–
gol–
imbol–
immel–
infil–
ingel–
insal–
invol–
merl–
modol–
niell–
orl–
pel–
pill–
popol–
regal–
rivel–
rotol–
rucol–
smel–
sparl–
spul–
still–
titol–
tutel–
util–

m-

abram–
acesm–
addom–
affam–
azzim–
bitum–
calm–
cim–
ciurm–
clam–
decim–
ferm–
gram–
illum–
innom–
inram–
inum–
letam–
mamm–

Roots of groups 11—20 *Radicaux des groupes 11—20*

menom-
pium-
ridam-
rilim-
rinom-
rugum-
rum-
spalm-
strem-
vidim-

n-

abbon-
adon-
adugn-
affun-
agogn-
agun-
amen-
argan-
asson-
bagn-
bulin-
capon-
caren-
cenn-
cign-
comun-
corn-
cucin-
dann-
degn-
desin-
deton-
dipan-
diren-
divin-
esorn-
etern-
faon-
infun-
latin-
macin-
marn-
nomin-
pern-
pian-
pugn-
raun-
ren-
ridon-
rimen-
riorn-
risan-
rovin-
scarn-
semin-
sogn-
spegn-
spin-
storn-
suon-
uman-
uncin-
velen-
zinn-
zon-

p-

capp-
colp-
crep-
dirup-
frapp-
grupp-
impep-
nasp-
pip-
popp-
ramp-
ricap-
scerp-
scip-
siep-
stamp-
stirp-
tarp-
zepp-

r-

abjur-
adir-
aggir-
alber-
allir-
ambr-
ammir-
arbor-
asper-
avver-
barr-
cedr-
cicur-
cifr-
coltr-
compr-
curr-
decor-
depur-
disir-
divor-
enarr-
esecr-
espir-
feltr-
figur-
foder-
giur-
immor-
impur-
inaur-
incar-
indor-
inter-
irror-
lacer-
lastr-
lavor-
libr-
logor-
mirr-
misur-
mostr-
mur-
natur-
nover-
onor-
oscur-
ottur-
peror-
riar-
rifor-
rigir-
rimur-
sacr-
sapor-
scor-
secur-
sevr-
smir-
sopr-
tempr-
tesor-
valor-
var-

s-

abiss-
abus-
accas-
annus-
arnes-
buss-
cans-
class-
disus-
fiss-
glos-
grass-
imbas-
invas-
iscus-
malus-
mass-
paves-
ras-
ripos-
sors-
spens-
spos-
tos-
vess-

t-

abort-
acet-
adast-
addit-
ajut-
amatt-
assit-
butt-
carat-
catt-
chet-
coort-
crost-
cuit-
dinot-
diput-
disat-
ditt-
eccit-
erutt-
fet-
fiut-
frett-
frust-
furt-
giunt-
gratt-
immit-
inset-
irret-
levit-
lott-
merit-
monet- 11–20
nett-
not-
onest-
ospit-
ovatt-
piot-
prest-
pront-
punt-
quart-
quet-
riatt-
rilut-
salt-
scult-
sest-
stut-

Roots of groups 11—20 *Radicaux des groupes 11—20*

tacit–
tast–
tent–
trit–
umett-
valut–
vant-
viet–
voit–
vuot–

u-

acqu-
adaqu–
adegu–
attu–
censu–
fru–
inaqu–
indu–
mutu–
situ–

v-

adjuv–
annev–
arriv–
attiv–
brav–
giov-
innav-
liev–
stiv–
trov–

z-

adizz–
agazz–
agenz–
avanz–
bozz–
cazz–
garz–
guazz–
olezz–
pinz–
pranz–
prezz–
rialz–
rizz–
spazz–
staz–
stizz-
terz–
unizz-

Group 21 (Italian)

refer to / *voir* } table 26 { see the following roots alphabetically arranged by their last letter / *voir les radicaux suivants rangés par ordre alphabétique de leur dernière lettre*

b-
cib-
rub-
turb-

d-
bard-
ered-
fid-
guid-
laid-
lord-
mund-
sald-
sod-
tond-

f-
beff-
grif-
olf-
stuf-
zaff-

j-
baj-
noj-
scoj-

l-
biul-
coll-
dobl-
isol-
zel-

m-
adim-
bram-
dam-
esum-
inam-
lem-
orm-
somm-
trem-

n-
eman-
magn-
pann-
san-
segn-
trin-
vign-

p-
cepp-
rap-
sarp-
topp-
zamp-

r-
cerr-
cur-
esor-
inor-
livr-
mitr-
odor-
oltr-
ombr-
sagr-
spar-
stir-

s-
adus-
bus-
dos-
gess-
mans-
nas-
pens-
scus-
spes-
stas-
tass-
vers-

t-
abit-
adot-
caut-
dett-
esit-
gitt-
latt-
lut-
mert-
pest-
pot-
quot-
rutt-
sost-
sput-
trot-
urt-
vent-
volt-

u-
equ-

v-
prov-
scav-

z-
alz-
forz-
ganz-
lenz-
marz-
mozz-
ponz-
suzz-

21

Groups 22—27 (Spanish)

Group **22**:	refer to / *voir* tables	17, 20/2, 29—33	see the following roots alphabetically arranged by their last letter
„ **23**:	„ „	17, 20/2, 28-33, 36	
„ **24**:	„ „	28, 36	
„ **25**:	„ „	17, 20/2, 29-30, 32/3	*voir les radicaux suivants rangés par ordre alphabétique de leur dernière lettre*
„ **26**:	„ „	20/2, 28—33, 36	
„ **27**:	„ „	20/2, 29—33	

Roots of groups 22—27 *Radicaux des groupes 22—27*

b-

adob–
aglob–
asorb–
cohob–
comb–
crib–
cub–
enalb–
engib–
enjeb–
estib–
galib–
garb–
gerib–
gib–
herb–
incub–
jorob–
silb–
sob-
trab–
tumb–

c-

apac–
busc–
cerc–
falc–
hinc-
pesc–

d-

abord–
aced–
adeud–
afond–
alaud–
alind–
alod-
anud–
apand–
arrid–
atild–
bald–
bebd–
boved–
brind–
camod–
cand-
coed–
cuid–
demud–
deved–
dubd–
embud-
enhad-
escud–
espad-
guard–
gueld–
guind–
had-
hered–
horad-
inund–
leud–
marid–
med–
moned–
olvid–
oxid–
rend–
sond–
told–

e-

ape–
fabe–

f-

afof–
alif–
arruf–
atuf–
bef–
chaf–
embaf–
engaf–
estof–
truf–

g-

ahog–
drag–
fisg–
freg–
pleg–

h-

acich–
agach–
ahuch–
atoch–
boch–
buh–
corch–
flech–
jach–
luch–
manch–
pinch–
tech–

i-

acuci–
agobi-
alaci–
alipi–
asoci–
atedi–
cambi–
dabi–
enri–
esfri–
expi–
fili-
forci–
inici–
istri–
jadi–
limpi–
lisi–
ludi–
murci–
nervi–
obvi–
odi–
ofici–
pali–
preci–
ripi–
roci–
siti–
uvi-
vaci–
vigi–

j-

abaj–
aduj–
afloj-
aguij-
ahij–
alej–
aquej–
arroj-
baraj–
batoj-
cobij-
cuaj-
dej–
desoj–
dibuj-
elij–
embij-
empaj–
encaj-
enej–
espej–
faj-
forj-
franj–
invej–

Roots of groups 22—27 *Radicaux des groupes 22—27*

manej-
moj-
motej-
puj-
tarj-
zanj-

l-

acopl-
afibl-
ahull-
anhil-
anill-
anubl-
apell-
arel-
argil-
asil-
atabl-
atipl-
azul-
birl-
bosl-
bull-
buril-
call-
camel-
cavil-
cicl-
copel-
corl-
empil-
enfil-
esmol-
fabul-
freil-
fusil-
habl-
inhal-
inmol-
jabel-
junel-
lill-
maul-
mell-
model-
mutil-
niel-
novel-
ojal-
ondul-
opil-
orill-
regl-
roll-
sopl-
sufl-

m-

abism-
abrom-
achin-
aclam-
adem-
afum-
agram-
ahirm-
alarm-
algam-
amim-
bizm-
blasm-
brum-
carem-
chom-
colm-
crism-
dezm-
difam-
encim-
engom-
enlam-
erm-
esfum-
exhum-
fem-
firm-
gorm-
husm-
jam-
lim-
pasm-
plom-
quem-
tom-

n-

abon-
aduan-
afin-
ahorn-
ajen-
almen-
alun-
amain-
aplan-
ardin-
arpon-
asign-
avan-
betun-
calon-
cecin-
churn-
copin-
dan-
desun-
devan-
disen-
domen-
empun-
endon-
enhen-
enrun-
ensan-
envin-
exorn-
fogn-
gran-
hacin-
iman-
lamin-
len-
maran-
mazn-
opugn-
orden-
orin-
pavon-
pein-
ruin-
sain-
tizn-
torn-

o-

apro-
barlo-
conlo-
inco-
mago-

p-

afelp-
agrup-
alamp-
apip-
arrap-
atop-
avisp-
charp-
chup-
disip-
dup-
empop-
enasp-
encep-
ensop-
equip-
escap-
guip-
jarop-
ocup-
rasp-
tap-
trep-
zarp-

r-

abebr-
abur-
acerr-
adjur-
admir-
adver-
ahetr-
ajor-
aladr-
alegr-
amer-
argur-
atair-
atigr-
aupr-
azorr-
azufr-
borr-
chipr-
chorr-
cobr-
demor-
edr-
entr-
esmer-
execr-
fatur-
ferr-
filtr-
honor-
jambr-
labr-
lembr-
letr-
logr-
madur-
mayor-
menor-
obtur-
oper-
peor-
pinor-
rear-
tir-
zurr-

s-

acens-
acos-
agris-
ahus-

22–27

Roots of groups 22—27 *Radicaux des groupes 22—27*

amas-
ascus-
aspas-
bras-
caus-
clis-
comis-
curs-
dehes-
depos-
devis-
drans-
enlos-
entes-
fos-
fris-
ofens-
opres-
prens-
sis-
soas-
ters-

t-

abrut-
abult-
acept-
acit-
adiet-
afect-
ajust-
alert-
alist-
aloit-
amant-
amort-
apart-
apest-
apret-
apunt-
asalt-
atet-
azot-
bolet-
brot-
capt-
chant-
cint-
cogit-
coyt-
cuart-
cuest-
enart-
enhot-
enjut-
escot-
espit-
evit-
exalt-
exent-
fart-
flet-
frut-
gust-
hurt-
imput-
inmut-
inst-
irrit-
jet-
junt-
lact-
mont-
musit-
objet-
ocult-
paut-
pect-
pist-
rest-
vot-

u-

actu-
adecu-
ajegu-
arcu-
atenu-
egu-
enagu-
gradu-
gru-
laqu-
licu-
mingu-
puntu-
valu-

v-

abrev-
activ-
adarv-
aviv-
azolv-
clav-
dadiv-
innov-
malv-
oliv-

x-

anex-
beix-
burx-
delax-

y-

apoy-
boy-
gay-

z-

aguz-
caz-
cruz-
driz-
lanz-

Group 28 (Spanish)

refer to / *voir* } tables 2, 27 { see the following roots alphabetically arranged by their last letter / *voir les radicaux suivants rangés par ordre alphabétique de leur dernière lettre*

Roots of group 28 *Radicaux du groupe 28*

b-

adub-
ajeb-
asob-
ceb-
grab-

d-

afod-
beld-
bond-
brid-
card-
fad-
feud-
fund-
mud-
sed-
tild-
vend-

f-

aruf-
orf-
pif-
zaf-

h-

cuch-
fech-
moch-
sach-

i-

agri-
bali-
cudi-
feri-
foli-
lidi-
vici-

j-

adoj-
aguj-
alij-
enoj-
fij-

l-

afil-
agol-
anul-
dol-
evol-
hall-
mirl-
moll-
nubl-
quel-
robl-

m-

aerm-
afam-
alom-
asum-
azem-
brom-
chum-
frem-
fum-
gom-
merm-
mim-
palm-
ream-

n-

acun-
asin-
aven-
bin-
born-
crin-
gan-
guin-
hen-
pun-
sign-

p-

apup-
lep-

r-

acer-
ator-
chor-
jorr-
jur-
lacr-
legr-
redr-
turr-
vir-

s-

ases-
atus-
avis-
cabs-
chas-
iris-
los-
mes-

t-

adat-
ahit-
anot-
cort-
frot-
gast-
hart-
list-
pact-
quit-
rent-
test-
tint-
trat-
vilt-

v-

ativ-
calv-

y-

ahoy-
moy-

28

29

Group 29 (Italian)

refer to / *voir* — table 34

see the following roots alphabetically arranged by their last letter

voir les radicaux suivants rangés par ordre alphabétique de leur dernière lettre

b-

esib–
forb–
inib–

c-

cusc–
dolc–
fulc–
sanc–

d-

alid–
cond–
fed–
lald–
sbad–
sped–
trad–

e-

inve–

g-

fugg–
larg–
sfug–

l-

abol–
fogl–
pol–

m-

grem–
vom–

n-

brun–
fin–
mun–
riun–

p-

carp–
comp–

r-

aper–
covr–
fior–
guar–
nodr–
nutr–
paur–
scur–
serr–

t-

acut–
cest–
malt–
ont–
sent–
tort–

u-

annu–
minu–
relu–
stru–

v-

civ–

z-

graz–

Bei Fragen zur Produktsicherheit wenden Sie sich bitte an:
If you have any questions regarding product safety,
please contact:

Walter de Gruyter GmbH
Genthiner Straße 13
10785 Berlin
productsafety@degruyterbrill.com